Synthese Library

Studies in Epistemology, Logic, Methodology, and Philosophy of Science

Volume 426

The aim of *Synthese Library* is to provide a forum for the best current work in the methodology and philosophy of science and in epistemology. A wide variety of different approaches have traditionally been represented in the Library, and every effort is made to maintain this variety, not for its own sake, but because we believe that there are many fruitful and illuminating approaches to the philosophy of science and related disciplines.

Special attention is paid to methodological studies which illustrate the interplay of empirical and philosophical viewpoints and to contributions to the formal (logical, set-theoretical, mathematical, information-theoretical, decision-theoretical, etc.) methodology of empirical sciences. Likewise, the applications of logical methods to epistemology as well as philosophically and methodologically relevant studies in logic are strongly encouraged. The emphasis on logic will be tempered by interest in the psychological, historical, and sociological aspects of science.

Besides monographs *Synthese Library* publishes thematically unified anthologies and edited volumes with a well-defined topical focus inside the aim and scope of the book series. The contributions in the volumes are expected to be focused and structurally organized in accordance with the central theme(s), and should be tied together by an extensive editorial introduction or set of introductions if the volume is divided into parts. An extensive bibliography and index are mandatory.

More information about this series at http://www.springer.com/series/6607

John Alexander Pinkston

Evidence and Hypothesis in Clinical Medical Science

 Springer

John Alexander Pinkston
Birmingham Radiological Group
Birmingham, AL, USA

Synthese Library
ISBN 978-3-030-44272-9   ISBN 978-3-030-44270-5   (eBook)
https://doi.org/10.1007/978-3-030-44270-5

This Springer imprint is published by the registered company Springer Nature Switzerland AG.
The registered company address is: Gewerbestrasse 11, 6330 Cham, Switzerland

In memory of my mother, Doris Jernigan Pinkston

Preface

I knew from an early age that I wanted to be a doctor, but it was during a humanities course as an undergraduate while otherwise immersed in premedical studies that I got hooked on philosophy. I went on to major in philosophy, but I had to temporarily leave formal study of the subject behind as I was off to medical school. This was followed by a period of military service, later to specialty training in radiation oncology, then to a career mostly as a radiation oncologist. Along the way, I got interested in epidemiology and statistics and did research and graduate work in those subjects, where it became apparent to me that philosophical concerns lie just beneath the surface of many problems that are approached in those fields, as well as in the broader area of clinical medicine itself. Only later would I find the opportunity to return to the formal study of philosophy and complete a doctoral program.

This book had its origins in a doctoral dissertation, and I am indebted to the many mentors whose influence has in some way made a contribution. Unfortunately, they are too numerous to mention here because they span my entire career. Special thanks, however, must go to those most directly responsible for helping me to see this project through to completion, and therefore, I wish to here acknowledge the help of Otávio Bueno, Edward Erwin, Kenneth Goodman, Ray Moseley, and Harvey Siegel.

In illustrating how evidence is gathered and used in the confirmation of the types of hypotheses that are commonly formulated in clinical medical science, and thus serving as background in my explication and defense of an approach to evidence that I am calling a "weight of evidence" account, I have made liberal use of examples and sources in the clinical medical scientific and philosophical literature. I would here like to express my appreciation to the numerous publishers and other copyright holders who have graciously given me permission to use previously published material. I have made every effort to properly credit or cite all sources used.

There have been instances where we have been unable to trace or contact the copyright holders. If notified, the publisher will be pleased to rectify any errors or omissions at the earliest opportunity.

Any mistakes in this book are entirely my own.

Birmingham, AL, USA   John Alexander Pinkston

Contents

Abbreviations

5-FU	5-fluorouracil
AE	atheromatous embolism
ATN	acute tubular necrosis
BCP	birth control pill
C.I.	confidence interval
CH50	complement
CMV	cytomegalovirus
EBM	evidence-based medicine
ECMO	extracorporeal membrane oxygenation
FARF	functional acute renal failure from dehydration
GN	glomerulonephritis
GTR	general theory of relativity
HIV	human immunodeficiency virus
HR	hazard ratio
IARC	International Agency for Research on Cancer
IN	interstitial nephritis
IRB	institutional review board
N of 1	sample size of one
NCCN	National Comprehensive Cancer Network
NIH	National Institutes of Health
NSABP	National Surgical Adjuvant Breast and Bowel Project
PSA	prostate-specific antigen
RCT	randomized controlled trial
TB	tuberculosis

Chapter 1
Introduction

Abstract This Chapter is the introduction. Here I lay out the principal aims of the book, which are to investigate the way that evidence is gathered and used to confirm the kinds of hypotheses that are commonly found in clinical medical science, and to introduce and defend a new theory that I call the "weight of evidence" account. I present some notions that have been offered of what we should expect from a theory of evidence. In the book I discuss five theories of evidence that have been proposed, namely hypothetico-deductivism, Bayesianism, Carl Hempel's "satisfaction" theory, Deborah Mayo's "error-statistical" theory, and Peter Achinstein's theory. I also consider Inference to the Best Explanation to the extent that it can be considered a method of theory choice. I advance reasons why I do not believe that previous theories satisfactorily explain confirmation in clinical medical science. I briefly outline the contents of the other chapters.

1.1 Aims and Motivation

My aim in what follows is to investigate the way that evidence is gathered and used to confirm hypotheses in clinical medical science. I use the term "clinical medical science" to include studies of both individual persons and groups of people where the subject matter concerns health and disease. Such studies may be centered in physicians' offices, clinics, hospitals, health departments, and the like. I consider three types of hypotheses that are commonly formulated: *therapeutic, etiologic, and diagnostic*. Therapeutic hypotheses are those that are concerned with treatments or other interventions; etiologic hypotheses are those concerned with the causes of disease; and, diagnostic hypotheses are those entertained by clinicians when making a diagnosis. How is evidence gathered and used to confirm these hypotheses in clinical medical science? My thesis is that a new account of evidence is needed to satisfactorily explain these processes, and another aim is to introduce what I am calling the *weight of evidence* account.

© Springer Nature Switzerland AG 2020
J. A. Pinkston, *Evidence and Hypothesis in Clinical Medical Science*, Synthese Library 426, https://doi.org/10.1007/978-3-030-44270-5_1

Several theories of evidence (or confirmation)[1] have been proposed, and while I have made no attempt to be exhaustive in examining them, to me the most salient ideas are to be found in five accounts that enjoy some degree of currency: Bayesianism, Karl Popper's elaboration of hypothetico-deductivism (1992), Carl Hempel's "satisfaction" theory (1945a, b), Deborah Mayo's "error-statistical" theory (1996), and Peter Achinstein's theory (2001, 2005). In these theories are to be found new approaches and efforts to improve on other accounts, and thus I believe they merit the attention that I will give them. I also consider "Inference to the Best Explanation," and address the extent to which explanatory factors should be considered in confirmation, since Achinstein has incorporated the requirement of an *explanatory connection* between a hypothesis and the evidence supporting it into his theory.

The present study is motivated by what seems to me to be inadequacies in these theories of evidence currently discussed in the philosophy of science. While each of the theories listed above, perhaps with the exception of Hempel's theory, can be considered to capture some of the important elements in the way that evidence is gathered and used in clinical medical science, none does so in a comprehensive and generally satisfactory way, or so I shall argue. The present effort is directed toward meeting these shortcomings.

What should we expect from a theory of confirmation? According to Clark Glymour,

> The aim of confirmation theory is to provide a true account of the principles that guide scientific argument insofar as that argument is not, and does not purport to be, of a deductive kind. A confirmation theory should serve as a critical and explanatory instrument quite as much as do theories of deductive inference. Any successful confirmation theory should, for example, reveal the structure and fallacies, if any, in Newton's argument for universal gravitation, in nineteenth-century arguments for and against the atomic theory, in Freud's arguments for psychoanalytic generalizations . . .
>
> The aim of confirmation theory ought not to be simply to provide precise replacements for informal methodological notions, that is, explications of them. It ought to do more; in particular, confirmation theory ought to *explain* both methodological truisms and particular judgments that have occurred within the history of science. By "explain" I mean that at least that confirmation theory ought to provide a rationale for methodological truisms and ought to reveal some systematic connections among them and, further, ought without arbitrary or question-begging assumptions to reveal particular historical judgments as in conformity with its principles (1980, 63–64).

Achinstein is more succinct:

> A theory of evidence, as I understand that expression, provides conditions for the truth of claims of the form
>
> (I) e is evidence that h,

---

[1] I use the expressions "theory of evidence" and "theory of confirmation" interchangeably (e.g., see Achinstein 1983, 351).

where e is a sentence describing some state of affairs and h is a hypothesis for which e provides the putative evidence (2005, 35).[2]

Glymour's notion is the more sweeping, and if he is correct, then ideally it would seem that a satisfactory theory of confirmation would be a comprehensive explanatory instrument for the inductive, empirical sciences. As such, it should be able to satisfactorily explain how evidence is gathered and used to confirm the variety of hypotheses encountered in clinical medical science. These include the kinds of hypotheses found in experimental studies such as randomized controlled trials (RCTs), non-experimental (usually observational) studies such as those concerning disease etiology, and those involved in making a diagnosis. Since I argue that none of the current theories satisfactorily does this, it follows that more work is needed.

Another reason why a new account of evidence may be of value is the frequency of "medical reversals," which has recently received increased attention (Prasad et al. 2013). Medical reversals are cases in which current or recent medical practice has been found to be inferior to some lesser or prior standard of care. They occur when new, better studies contradict current practice that is based on prior, inferior evidence. An example is coronary artery stenting. For decades, stenting of the coronary arteries in the initial management of stable coronary artery disease was very common, but the practice was contradicted when it was found to be no better than optimal medical management alone for most patients (Boden et al. 2007). Prasad et al. analyzed 146 such instances of medical reversals, which they find to be common and to occur across all classes of medical practice. Thus a new account of evidence focused on the types of hypotheses found in clinical medical science and how they are confirmed, with attention to sources of error, seems especially timely.

1.2   Overview of the Following Chapters

The current theories of confirmation that I will discuss can be broadly divided into two groups: those that consider hypotheses to have probabilities (i.e., p(h) = r ; 0 ≤ r ≤ 1), and those that do not. The theories of Popper, Hempel, and Mayo fall into the latter group, whereas Bayesianism and Achinstein's theory fall into the former. In Chap. 2, I will discuss the theories of Popper, Hempel, and Mayo, including what I believe to be strengths and weaknesses along with some objections to these theories that have been offered by others.

In Chap. 3, I treat Bayesianism and Achinstein's theory in much the same way as I do the theories in Chap. 2, and in addition I consider Inference to the Best Explanation and Achinstein's "explanatory connection" requirement in the context

---

[2]In a note, Achinstein states that more accurately, it is the fact that e, rather than the sentence describing that fact, that constitutes the evidence, but that he is following standard philosophical practice of speaking of the sentence e.

of the role of explanation in clinical medical science. I will argue that explanation[3] is not necessary for confirmation.

In Chap. 4, I consider therapeutic, etiologic, and diagnostic hypotheses, and illustrate and discuss each type with one or more examples drawn from the scientific medical literature. An RCT and an N of 1[4] study are presented as examples for evaluating therapeutic hypotheses, and cohort, case-control, and cross-sectional studies are used as examples for etiologic hypotheses. Diagnostic hypotheses are illustrated with cases drawn from published clinicopathologic conferences and similar sources, with an emphasis on the various strategies employed to arrive at a diagnosis.

In Chap. 5, I will explicate the weight of evidence account. My emphasis will be on the accuracy of individual observations and studies, where accuracy is understood to be a function of validity and precision. Observations and studies are accurate just to the extent that they are valid and precise. Weight of evidence is a function of accuracy.

I will defend the weight of evidence account in Chap. 6, where I argue that it remedies the deficiencies in the other accounts, and also satisfactorily explains the case studies, as well as explaining the various efforts to rank evidence.

The evidence-based medicine (EBM) movement is a current effort within the medical scientific community to place medicine on a sounder scientific evidentiary basis. Part of its approach is to rank sources of evidence according to the degree of confidence that should be placed in them. It is sometimes illustrated as a "hierarchical pyramid" with RCTs (or systematic reviews and meta-analyses of RCTs) at the apex, with case reports, expert opinion, and the like at the base. Cohort, case-control, and cross-sectional studies fall into the middle. In Chap. 7, I use the weight of evidence rationale to provide a philosophical justification for the EBM hierarchical ranking of studies and other sources of evidence. In addition, I defend the need for randomization in RCTs against critics from both the philosophical and medical communities. I also argue that the weight of evidence account explains the various "levels" of evidence encountered in clinical medical science, and illustrate this with the historical case study of the evolution of treatments for early breast cancer.

It will become clear that although historically many of the approaches used in medical practice are based on the experiences of clinicians, and have not evolved based on formal testing, in the modern era research questions in clinical medical science are answered by the formal gathering and using of evidence for or against hypotheses. Since this includes interventions that in some cases may involve serious risk, ethical questions inevitably arise. Are there limits to acquiring evidence in human beings? Can studies designed to improve therapeutic outcomes, or to decrease mortality or morbidity, be unethical? In Chap. 8, I will discuss these issues,

---

[3]By "explanation" here is meant what is ordinarily construed as explanation, i.e., that it is context-dependent and dependent on the particular interests of persons. It may not necessarily pertain to the notion of explanation used by Achinstein in his "explanatory connection," as I later argue.

[4]An N of 1 study is a study involving a single person.

and argue that it is possible to determine when sufficient evidence has been acquired to justify the use of a new therapy. I will argue that evidence from RCTs is not invariably necessary, and illustrate this by discussing the ethical issues surrounding a series of studies on respiratory therapies in neonates. I also describe how and why sufficient evidence accumulated to establish generally accepted definitive therapies in disseminated cancer of the testis and cancer of the anal canal without subjecting the hypotheses on which they were based to RCTs.

In the chapters that follow I have made extensive use of examples from the scientific medical literature. Although most are relatively recent, some are less so and may even date back several decades. These less recent examples were chosen because for the most part they are the best studied and least controversial, and many are considered classical. Since they are among the best known and most compelling, I believe they serve as among the best examples for the points I make.

References

Achinstein, Peter. 1983. *The nature of explanation*. Oxford: Oxford University Press.
———. 2001. *The book of evidence*. Oxford: Oxford University Press.
———. 2005. Four mistaken theses about evidence, and how to correct them. In *Scientific evidence. Philosophical theories and applications*, ed. Peter Achinstein, 35–50. Baltimore: Johns Hopkins University Press.
Boden, William E., Robert A. O'Rourke, Koon K. Teo, Pamela M. Hartigan, David J. Maron, William J. Kostuk, Merril Knudtson, et al. 2007. Optimal medical therapy with or without PCI for stable coronary disease. *New England Journal of Medicine* 356: 1503–1516.
Glymour, Clark. 1980. *Theory and evidence*. Princeton: Princeton University Press.
Hempel, Carl G. 1945a. Studies in the logic of confirmation (I). *Mind* 54: 1–26. Reprinted, with some changes, in Hempel 1965, 3–51
———. 1945b. Studies in the logic of confirmation (II). *Mind* 54: 97–121. Reprinted, with some changes, in Hempel 1965, 3–51.
———. 1965. *Aspects of scientific explanation and other essays in the philosophy of science*. New York: The Free Press.
Mayo, Deborah G. 1996. *Error and the growth of experimental knowledge*. Chicago: University of Chicago Press.
Popper, Karl. 1992. *The logic of scientific discovery*. New York: Routledge.
Prasad, Vinay, Andrae Vandross, Caitlin Toomey, Michael Cheung, Jason Rho, Steven Quinn, Satish Jacob Chacko, et al. 2013. A decade of reversal: An analysis of 146 contradicted medical practices. *Mayo Clinic Proceedings* 88: 790–798.

Chapter 2
Theories of Confirmation in Which Hypotheses Do Not Have Probabilities

Abstract In this chapter, I consider three theories of evidence, namely, hypothetico-deductivism, Carl Hempel's "satisfaction" theory, and Deborah Mayo's "error-statistical" theory. These theories are considered together because they share the characteristic that hypotheses themselves do not have probabilities (in contrast to those that do, i.e., p(h) = r; $0 \leq r \leq 1$, which are considered in Chap. 3). I discuss hypothetico-deductivism mainly using Karl Popper's approach to this theory. I attempt to explicate these theories in sufficient detail so that in later chapters I will be able to argue that they do not satisfactorily explain confirmation in clinical medical science, and that the "weight of evidence" account does this more satisfactorily. I also include some objections to these theories that have been offered by others.

2.1 Hypothetico-deductivism

Hypothetico-deductivism, which is sometimes called *falsificationism* or *refutationism*, was developed in considerable detail in an influential book by Karl Popper (1992).[1] Popper rejected inductivism as a method for the acquisition of scientific knowledge. He maintained that hypotheses, once formed, should be subjected to test in order to falsify them. For Popper, hypotheses are not confirmed by positive instances or observations consistent with or predicted by them; rather, they are only *corroborated* by such evidence and by withstanding attempts at falsification. For example, no matter how many white swans we observe, we are never justified in concluding that *all* swans are white (1992, 4). The extent to which a hypothesis has been subjected to testing in an attempt to falsify it is the extent to which it has "proved its mettle" (1992, 10). The logical form of Popper's method is (Butts 1995, 352–353; Popper 1992, 54–56):

---

[1]This book originally appeared in German in 1934. The first English translation was in 1959 with the title *The Logic of Scientific Discovery* (London: Hutchinson, 1959).

© Springer Nature Switzerland AG 2020

J. A. Pinkston, *Evidence and Hypothesis in Clinical Medical Science*, Synthese Library 426, https://doi.org/10.1007/978-3-030-44270-5_2

$$(H \cdot A) \rightarrow O$$
$$O$$
$$\overline{\therefore (H \cdot A)}$$

Here, H signifies a hypothesis that is conjoined with one or more auxiliary hypotheses or initial conditions A, and O denotes an observation. Thus, even if a hypothesis logically implies an observation O, and that observation is made, one cannot assume the truth of the hypothesis since infinitely many other hypotheses and their conjuncts could theoretically imply the same observation. It is deductively invalid, committing the logical fallacy of *affirming the consequent*. However, the attempt to refute (H · A) by making an observation that contradicts O (i.e., *not* O) would falsify the conjunct (H · A), and is deductively valid by *modus tollens*:

$$(H \cdot A) \rightarrow O$$
$$\sim O$$
$$\overline{\therefore \sim (H \cdot A)}$$

Popper regards scientific theories as universal statements (1992, 37). To give a causal explanation of an event is to deduce a statement that describes it, using as premises in the deduction one or more universal laws along with certain singular statements, the initial conditions or auxiliary hypotheses. For example, to explain the breaking of a thread that has a tensile strength of one pound when a weight of two pounds was placed on it might be constructed as follows: H: "Whenever a thread is loaded with a weight exceeding that which characterizes the tensile strength of the thread, then it will break." This statement has the character of a universal law of nature. Statements of initial conditions might be "The weight characteristic for this thread is 1 lb.", and, "The weight put on this thread was 2 lbs." From these statements we deduce the prediction that the thread will break. For Popper, the initial conditions describe what is usually called the *cause* of the event, and the prediction describes what is usually called the *effect*. However, Popper wishes to avoid the use of the terms "cause" and "effect", regarding any "principle of causality" (i.e., the assertion that any event whatever *can* be causally explained, and that it *can* be deductively predicted) as metaphysical. He does, however, propose the methodological rule that we should continue to search for universal laws and coherent theoretical systems, and seek causal explanations for the types of events that we can describe (1992, 38–39).

Popper distinguishes the process of conceiving a new idea, or the generation of hypotheses and theories, and the methods of examining them logically. For Popper, there is no such thing as a logical method of having new ideas, or any logical reconstruction of the process. Every discovery contains an "irrational" element or a "creative intuition". He writes, "In a similar way Einstein speaks of the 'search for those highly universal laws ... from which a picture of the world can be obtained by pure deduction. There is no logical path', he says, 'leading to these ... laws. They can only be reached by intuition, based upon something like an intellectual love ('*Einfühlung*') of the objects of experience'" (1992, 8–9).

For Popper, the postulate of the existence of universal laws of nature is an example of a methodological rule. Methodological rules are *conventions*. They are the rules of the game of science, much as the game of chess has rules (1992, 32). It is part of our definition of natural laws that they are invariant with respect to space and time, and derives from our faith in the "uniformity of nature," which is also a metaphysical concept (1992, 250–251).

Popper notes that probability statements are impervious to strict falsification (1992, 133), and yet such statements as well as probabilistic hypotheses are extant in science. For example, although we may regard the hypothesis "This is a fair coin" as falsified if it invariably turns up heads, nevertheless the number of tosses is finite and there cannot be any question of falsification in a logical sense. Indeed, Popper says, "Probability hypotheses *do not rule out anything observable …*" (1992, 181). Yet Popper acknowledges the success science has had with predictions obtained from hypothetical estimates of probabilities, and proposes that probability hypotheses can play the role of natural laws in empirical science. Thus he proposes that we accept something like what he believes a physicist might offer as a practically applicable physical concept of probability:

> There are certain experiments which, even if carried out under controlled conditions, lead to varying results. In the case of some of these experiments – those which are 'chance-like', such as tosses of a coin – frequent repetition leads to results with relative frequencies which, upon further repetition, approximate more and more to some fixed value which we may call the *probability* of the event in question. This value is '… empirically determinable through long series of experiments to any degree of approximation';[2] which explains, incidentally, why it is possible to falsify a hypothetical estimate of probability. (1992, 191)

Popper rejects the notion that hypotheses can be "probably true" based on tests, since the idea that scientific theories can be justified or verified, or even probable, is based on induction. He regards as a mistake the historical idea that science is a body of knowledge that is progressing toward truth, but that once it became clear that certain truth was unattainable, perhaps that it could be considered as "probably true". He believes that "… we must not look upon science as a 'body of knowledge', but rather as a system of hypotheses; that is to say, as a system of guesses or anticipations which in principle cannot be justified, but with which we work as long as they stand up to tests, and of which we are never justified in saying that we know that they are 'true' or 'more or less certain' or even 'probable'" (1992, 318).

As Achinstein (2005, 3) has noted, Popper's view on falsification accords with the views of some scientists. For example, the physicist Richard Feynman writes:

> In general we look for a new law by the following process. First we guess it. Then we compute the consequences of the guess to see what would be implied if this law that we guessed is right. Then we compare the result of the computation to nature with experiment or

---

[2]In a note, Popper cites several sources for this quotation, which I do not repeat here.

experience, compare it directly with observation, to see if it works. If it disagrees with experiments it is wrong. In this simple statement is the key to science ...

You can see, of course, that with this method we can attempt to disprove any definite theory. If we have a definite theory, a real guess, from which we can conveniently compute consequences which can be compared with experiment, then in principle we can get rid of any theory. There is always the possibility of proving any definite theory wrong, but notice that we can never prove it right. Suppose that you invent a good guess, calculate the consequences, and discover every time that the consequences you have calculated agree with experiment. The theory is then right? No, it is simply not proved wrong (Feynman 1967, 156–157).

A principal objection to Popper's theory is that hypotheses are rarely if ever tested in isolation, but are accompanied by a variety of initial conditions or auxiliary hypotheses. Thus, suppose that a hypothesis, conjoined with several auxiliary assumptions, logically entails an observation sentence, and that the observation sentence is found to be false. Then the conjunction must be false, so at least one conjunct must be false, but where do we place the blame for the negated prediction (Glymour 1980 30–31; Earman 1992, 65)?

Hempel has also argued that Popper's theory of admitting only relatively falsifiable sentences is overly restrictive in that it severely limits the possible forms of scientific hypotheses. For example, it rules out all purely existential hypotheses as well as most hypotheses that require both universal and existential quantification, and thus is inadequate to explicate satisfactorily the status and function of more complex scientific theories and hypotheses (1945b, 119–120).

2.2 Hempel's "Satisfaction" Theory

Carl Hempel set out to formulate the basic principles of a logic of confirmation (1945a, b). He wanted to characterize in precise and general terms the conditions under which a body of evidence can be said to confirm or disconfirm an empirical hypothesis. Hempel sought *objective* criteria in which there was no necessary mention of the subject matter of the evidence or hypothesis; he wanted *formal* criteria of confirmation in the way that deductive logic provides formal criteria for the validity of deductive inferences. Confirmation can be construed as a relation between sentences: for example, "a is a raven & a is black" is an evidence sentence confirming the hypothesis sentence "All ravens are black". Hempel interprets confirmation as a logical relation between sentences (1945a, 22).

Evidence sentences are like observation reports, for example, direct observation (e.g., "black"), but not theoretical constructs, like "heavy hydrogen" (1945a, 22–23). "An observation sentence describes a possible outcome of the accepted observational techniques" (1945a, 24). An observation report can be a conjunction of sentences or a class of sentences (1945a, 25). Evidence is *relevant* to a hypothesis only if it tends to confirm or disconfirm it (1945a, 5).

For Hempel, "...an adequate analysis of scientific prediction (and analogously, of scientific explanation and of the testing of empirical hypotheses) requires an analysis of the concept of confirmation" (1945b, 101). Many scientific laws and theories connect terms that are theoretical constructs rather than those of direct observation. From observation sentences, no merely deductive logical inference leads to statements about theoretical constructs that can serve as a starting point for predictions; statements about theoretical constructs such as "This piece of iron is magnetic" can be confirmed, but not entailed, by observation reports. Thus, even though based on general scientific laws, the prediction of new observational findings by means of given ones is a process involving confirmation in addition to logical deduction (1945b, 101–102).

Hempel outlines several conditions of adequacy of any definition of confirmation (1945b, 102–105):

(1) Applicable to hypotheses of any degree of logical complexity, in addition to simple universal conditionals.
(2) If a sentence is entailed by an observation report, then it is confirmed by it (entailment condition).
(3) If an observation report confirms every one of some particular class of sentences, then it also confirms any sentence that is a logical consequence of that class (consequence condition).
(4) An observation report that confirms a hypothesis h also confirms every hypothesis that is logically equivalent to h (equivalence condition).
(5) Logically consistent observation reports are logically compatible with the class of all the hypotheses that they confirm (consistency condition).

While (1)–(5) are necessary, they are not sufficient. A definition of confirmation must also be "materially adequate" and "provide a reasonably close approximation to that conception of confirmation which is implicit in scientific procedure and methodological discussion." Also required is reference to some accepted "language of science" applicable to all observation reports and hypotheses being considered, and with a precisely determined logical structure (1945b, 107).

Hempel's *satisfaction criterion of confirmation* in essence states that a hypothesis is confirmed by some particular observation report if the hypothesis is satisfied in a finite class K of those individual members of K which are mentioned in the report; that is, an observation report confirms a hypothesis H if H is entailed by a class of sentences, each of which is directly confirmed by the observation report (1945b, 109). An observation report disconfirms a hypothesis H if it confirms the denial of H. And, an observation report is neutral with respect to H if it neither confirms nor disconfirms H (1945b, 110).

If an observation report entails H, then it conclusively confirms (verifies) H. If the report entails the denial of H, then it conclusively disconfirms (falsifies) H. These concepts of verification and falsification are *relative*; a hypothesis is verified or falsified only with respect to some observation report (1945b, 112). Absolute verification does not belong to logic, but rather to pragmatics: it refers to acceptance of H by scientists on the basis of relevant evidence (1945b, 114).

Hempel outlines three phases of scientific tests of hypotheses (1945b, 114):

(1) Performance of experiments or making observations with acceptance of the observation reports as relevant to H.
(2) Assessing H with regard to the accepted observation reports, which confirm or disconfirm H, etc.
(3) Accepting or rejecting H based on the observation reports, or suspending judgment, or awaiting further evidence, etc.

Hempel is mainly concerned with (2), which for him is purely logical, and invokes only logical concepts. (1) and (3) are pragmatic; for example, (3) usually is tentative and could be changed (1945b, 114–115). Relative verifiability or falsifiability is a simple logical fact, but absolute verifiability or falsifiability may not be attainable in empirical science since new evidence or hypotheses can overturn any previous evidence or hypotheses (1945b, 116–119).

Statistical syllogisms, unlike deductive methods, can lead to inconsistencies, i.e., incompatible conclusions. Hempel (1960, 440) notes that Toulmin (1958, 109) has put forward as valid certain types of argument that he calls "quasi-syllogisms." These can take forms such as the following:

a is F
The proportion of F's that are G is less than 2%
So, almost certainly (or: probably), a is not G[3]

Consider the following argument (Hempel 1960, 441; Toulmin 1958, 109):

Petersen is a Swede
The proportion of Roman Catholic Swedes is less that 2%
So, almost certainly, Petersen is not a Roman Catholic

Suppose that the premises are true. But, as Cooley (1959, 305) notes, the following can also be true:

Petersen made a pilgrimage to Lourdes
Less that 2% of those making a pilgrimage to Lourdes are not Roman Catholics
So, almost certainly, Petersen is a Roman Catholic[4]

The conclusions are incompatible, but with a deductive syllogism this cannot happen:

a is F
All F are G
a is G

---

[3]For the conclusion form "almost certainly, or probably, a is not G," see Toulmin (1958, 139).

[4]Hempel (1960, 441) notes that he has slightly modified the phrasing of Cooley's example to more closely fit the pattern of the other examples.

If the premises are true, there is no rival argument of the same form whose premises are true as well and whose conclusion is incompatible with that of the given argument. Incompatible conclusions can only be deduced from incompatible sets of premises, and sets of true premises are not incompatible (Hempel 1960, 443).

Thus, in considering what to do when, for example, we have two sets of premises that are true but the conclusions are incompatible for, say, some unknown state of affairs or future event, Hempel maintains that we should use the *totality* of the available evidence. Quoting Carnap, he says: "In the application of inductive logic to a given knowledge situation, the total evidence available must be taken as a basis for determining the degree of confirmation" (Hempel 1960, 451).[5] This principle, according to Hempel, specifies a necessary, though not sufficient, condition for the rationality of inductive beliefs and decisions. Accepting certain statements, like the notion of "total evidence", is pragmatic (1960, 453).

Hempel's account has been criticized on several grounds (Earman 1992, 68–69). For example, it is silent on how theoretical hypotheses are confirmed, since evidence statements are in a purely observational vocabulary.

It has also been criticized as too liberal, since, for example, it allows confirmation of Goodman's "new riddle" cases (Goodman 1983). Consider: "All ravens are black": $(\forall x)\ (Rx \to Bx)$. Let x be not black but blite: i.e., black if examined before 2000 and white if examined after. On Hempel's account it confirms the prediction that after 2000 ravens will be white if up until then they have been black.

Earman (1992, 74) notes that any account of confirmation modeled on Hempel's approach will have two major problems: (1) For Hempel, whether evidence confirms a hypothesis depends only on the syntax, but we know from the Goodman example above that this is wrong. (2) For Hempel, confirmation is a 2-place relation, but we know that background information must be brought into the analysis to get an illuminating treatment. Earman provides a couple of examples. Consider the first (which I paraphrase):

H1: "All ravens are black" $(\forall x)\ (Rx \to Bx)$
H2: "All ravens live happily in heaven after they die" $(\forall x)\ (Rx \to Hx)$

Suppose that the two hypotheses H1 and H2 constitute theory T that is logically closed. Thus, it is part of T that H3: $[Rx \to (Bx \leftrightarrow Hx)]$. So from a confirming instance of H1, say, Ra & Ba, we can deduce via H3 that Ra & Ha, which is a Hempel positive instance of H2, which is absurd. However, consider the second example:

H1: "All patients with symptoms S have antibodies to a certain virus" $(\forall x)$ $(Sx \to Ax)$
H2: "All patients with symptoms S are infected with said virus" $(\forall x)\ (Sx \to Vx)$

If the evidence is Sa & Aa, it seems perfectly reasonable to deduce Sa & Va. But structurally, syntactically, the examples are the same. The difference is in the

---

[5]Hempel cites several references from Rudolf Carnap for this quotation, which I do not repeat here.

background information (74–75).[6] Others have also stressed that confirmation is a 3-place relation between data, the hypothesis in question, and background information (e.g., Howson and Urbach 2006, 298–299).

2.3 Mayo's Error-Statistical Theory

Deborah Mayo (1996) presents and elaborates in detail her error-statistical theory of confirmation, which rests on the notion that for a hypothesis to be confirmed it must be subjected to and pass a *severe test*. Thus, evidence e should be taken as good grounds for hypothesis H to the extent that H has passed a severe test with e. A passing result is a severe test of the hypothesis H just to the extent that it is very improbable for such a passing test to occur, were H false. Her theory largely rests on concepts in probability theory and (classical) statistics, and does not require that probabilities be assigned to hypotheses.

Requirements for error severity are (Mayo 1996, 178–180):

(1) e must "fit" H
(2) e's fitting H must constitute a good test of H. It rules out "e is a poor test of H" and at the same time "e is evidence for H."
(3) Severity criterion (for experimental testing contexts)

  (a) There is a very high probability that test procedure T would *not* yield such a passing result, if H is false
  (b) There is a very low probability that test procedure T would yield such a passing result, if H is false

Mayo (1996, 182–183) offers examples of minimum and maximum severity tests:

Minimum severity (zero-severity) test
H passes a zero-severity test with e if and only if test T would always yield such a passing result even if H is false
H: Student can recite aloud the name of all 50 U.S. state capitals
T: Student can recite *anything* aloud, ∴ evidence e
Result: Student always passes

Maximum severity (100% severity) test
H passes a maximally severe test with e if and only if test T would never yield results that accord with H as well as e does, if H is false
H: Student can recite aloud the name of all 50 U.S. state capitals
T: Test student, e = student can recite aloud the name of all 50 capitals

Mayo's emphasis is on the detection and elimination of error in scientific experimentation. For Mayo, most progress in science and the growth of experimental

---

[6]Earman cites Christensen (1983, 1990) as references for the two examples.

knowledge occurs in the quotidian scientific laboratory or other experimental venue, and is most akin to Kuhn's "normal science" (Mayo 1996, 26–31) or Hacking's "topical hypotheses" (Mayo 1996, 61). Her idea is rather than "go big," like Einstein's General Theory of Relativity (GTR), we should "go small," and focus on local hypotheses. Normal experimental testing is the testing of local hypotheses, and is, for Mayo, what justifies her heavy reliance on frequentist (classical) statistics (1996, 459).

Mayo (1996, 18) draws attention to several sources of error, including the design and control of experiments, and how data are generated, modeled, and analyzed. Four canonical or standard sources of error that must be addressed are:

(1) Mistaking experimental artifacts for real effects; mistaking chance effects for genuine correlations or regularities
(2) Mistakes about a quantity or value of a parameter
(3) Mistakes about a causal factor
(4) Mistakes about the assumptions of experimental data

These sources of error are addressed by methodological rules, including rules for pretrial planning, conduct of experiments, and post experiment analysis. The rules are empirical claims or hypotheses about how to find out things from experiments, and about how to proceed in given contexts to learn from experiments. Thus, she contends that her model of an epistemology of experiment is both naturalistic and normative.

According to Mayo (1996, 128–129), an adequate account of experimental testing must not begin at the point where data and hypotheses are given, but also must incorporate the intermediate theories of data, instruments, and experiment that are required to obtain the experimental data in the first place. At least three models are involved:

(1) *Primary scientific hypotheses or questions.* A substantive scientific inquiry is to be broken down into one or more local or "topical" hypotheses, each of which corresponds to a distinct *primary question* or *primary problem.*
(2) *Experimental models.* The experimental models serve as the key linkage models connecting the primary model to the data, and conversely.
(3) *Data models.* Modeled data, not raw data, are linked to the experimental models.

Mayo (1996, 141–144) illustrates her approach by considering an RCT carried out in Puerto Rico (Fuertes-de la Haba et al. 1971) to determine if women taking birth control pills (BCPs) were at increased risk of thromboembolism. The *primary question* in this study was whether there is an increased risk of blood clotting disorder among women using BCPs (for a specified length of time). Hypothesis H: The incidence of clotting disorder in women taking BCPs does not exceed the incidence among control women (women not taking BCPs). The study was modeled as a difference in average incidence rates: $\Delta = \mu_T - \mu_C$, where μ_T = incidence rate among BCP users and μ_C = incidence rate among controls. The primary question was to test the null hypothesis of no difference in incidence rates H_o: $\Delta = 0$ vs. the alternative hypothesis H′: $\Delta > 0$.

The *experimental* model was to specify two groups, for example around 5000 women in each group. The observed rates in the two groups are represented by the means in the two samples, $\overline{X}_T$ and $\overline{X}_C$, where $\overline{X}_T$ and $\overline{X}_C$ represent the sample means for BCP users and nonusers, respectively. We define a risk increase RI for BCP users to be $\overline{X}_T - \overline{X}_C$. If H_o is true, we expect RI = 0. The distance from H_o = observed RI – expected RI. The further the distance, the less likely H_o is to be true. This may be reported as a significance level.

To generate the experimental data, approximately 10, 000 women were randomly assigned to each group. The recorded data, the *data* model, is the value of RI obtained once the experiment has been performed and the results analyzed. In utilizing the observed RI, it is important to see that other risk factors (factors other than BCPs that could spuriously alter the results) were also evenly distributed between the two groups. The RI in this experiment was 0.0002, consistent with the inference of no difference in incidence rates.

Another example of a canonical model of error is the binomial model of experiment, for example, coin toss experiments that can be modeled as success or failure. Let the probability of success of each trial = p (which are Bernoulli trials with parameter p), which can be modeled using the binomial distribution, with n trials. An example is the coin toss. Here, considering the coin to be a fair coin (i.e., the probability of heads = the probability of tails), p = 0.5. The probability of "success," (e.g., the relative frequency of heads), is given by:

$$P(\text{relative frequency of success} = k/n) = \frac{n!}{k!(n-k)!}p^k(1-p)^{n-k} \qquad (2.1)$$

To illustrate, consider the example of a lady tasting tea, who claims to be able to tell whether milk is added to the cup before or after the tea has been poured (Mayo 1996, 154–160).[7] The *primary* question is whether she can do better than mere guessing. Thus we have H_o: p = 0.5 (just guessing) versus the hypothesis H': p > 0.5, in which she does better than expected for someone just guessing. Suppose we let her taste 100 cups of tea. The *experimental* statistic is the relative frequency of successes in the n experimental trials, $\overline{X}$. Suppose we let n = 100. In the *data* model, we concern ourselves with ensuring that the experimental assumptions are met, for example, factors that might tip off the lady are eliminated (e.g., all the cups are alike, the same amounts of milk and tea in each cup, etc.). The tests constitute a single data set, and address a specific question. The test statistic here is the distance measure $D(\overline{X}) = \overline{X}_{\text{observed}} - \overline{X}_{\text{expected}}$. We will determine the statistical significance level of the difference D_{obs} (in testing H_o), which is the probability of obtaining a difference as large as or larger than D_{obs}, assuming H_o is true.

Suppose that the observed proportion is 0.6, i.e., she successfully identified 60 of the 100 tastings. Then: $P(\overline{X} \geq 0.6) = 0.03$. The frequentist statistical interpretation

---

[7]The tea tasting example is usually attributed to R.A. Fisher (1947). As Mayo notes, she has slightly modified Fisher's presentation.

here is that the probability that the lady would get at least 60 successes (or more) out of 100 tries if she were just guessing is 0.03. This is the probability of incorrectly rejecting the hypothesis H_o (which is the "null" hypothesis of just guessing) when in fact she *is* just guessing. This is called the α level of significance, or the probability of a Type I error. It is the probability of rejecting the null hypothesis when it is true. This trial constitutes for Mayo a *severe* test of the null hypothesis, and thus, whether to accept the hypothesis that she *can* tell the difference, since the probability of just guessing is so low (0.03). Mayo's *measure of severity* is $1 - \alpha$, which in this case is 0.97. Since α denotes a probability, the numerical value of $1 - \alpha$ will lie between 0 and 1.

The statistical theory of experiment deals only with certain kinds of experiments insofar as their results can be modeled by certain parameters. A characteristic of key interest is the relative frequency with which certain results obtain. We want to separate "signal" from "noise." We have "subtracted out" the effect of mere guessing in the tea lady example. Thus, for Mayo, the use of statistical methods is what allows her error-statistical theory to be "truly ampliative" (Mayo 1996, 443).

Mayo (1996, 278–292) also defends her theory as applicable to "big" theories like Einstein's GTR, to which the theory in the totality might be difficult or impossible to assign a frequentist interpretation, but nevertheless to which many of the local hypotheses entailed by the theory could be subjected to test in the piecemeal "normal science" methodology which she claims is how experimental knowledge grows. Thus, in the 1919 test of the bending of starlight by the sun, Einstein's theory predicted an almost doubling of the deflection at the limb of the sun as compared to Newtonian theory. Attention to multiple possible sources of error, along with the determination of point estimates with statistical margins of error recorded, Mayo believes accords well with her approach. The test results, considered a severe test by her criteria, were in line with Einstein's predictions and confirmed his theory.

A strength of Mayo's error-statistical theory is that the frequentist approach is widely used in the testing of statistical hypotheses in the sciences, including clinical medical science. Giora Hon (2003, 191–193), however, has criticized her theory as being too narrowly focused on error probabilities, and that her philosophy of experiment relies neither on scientific theories nor on a theory of experiment, but instead on methods – statistical methods – for producing experimental effects. He argues that her focus is primarily on statistical calculations rather than the actual practice of experimentation, and errors are not errors at large, but instead statistical error probabilities, and that while it can be said that her approach does constitute a contribution to experimental design in the traditional sense of the term as well as to an analysis of error probabilities, it does not illuminate the inner epistemic processes of experiments and no theory of experiment is forthcoming. Thus, for Hon, while Mayo (1996, 444) claims that her theory is a "full-bodied experimental philosophy", it comes up short.

References

Achinstein, Peter. 2005. Four mistaken theses about evidence, and how to correct them. In *Scientific evidence. Philosophical theories and applications*, ed. Peter Achinstein, 35–50. Baltimore: Johns Hopkins University Press.

Butts, Robert E. 1995. Hypothetico-deductive method. In *The Cambridge dictionary of philosophy*, ed. Robert Audi, 352–353. Cambridge: Cambridge University Press.

Christensen, David. 1983. Glymour on evidential relevance. *Philosophy of Science* 50: 471–481.

———. 1990. The irrelevance of bootstrapping. *Philosophy of Science* 57: 644–662.

Cooley, J.C. 1959. On Mr. Toulmin's revolution in logic. *The Journal of Philosophy* 56: 297–319.

Earman, John. 1992. *Bayes or bust? A critical examination of Bayesian confirmation theory.* Cambridge: MIT Press.

Feynman, Richard. 1967. *The character of physical law.* Cambridge: MIT Press.

Fisher, Ronald A. 1947. *The design of experiments.* 4th ed. Edinburgh: Oliver and Boyd.

Fuertes-de la Haba, Abelardo, José O. Curet, Ivan Pelegrina, and Ishver Bangdiwala. 1971. Thrombophlebitis among oral and nonoral contraceptive users. *Obstetrics and Gynecology* 38: 259–263.

Glymour, Clark. 1980. *Theory and evidence.* Princeton: Princeton University Press.

Goodman, Nelson. 1983. *Fact, fiction, and forecast.* 4th ed. Cambridge: Harvard University Press.

Hempel, Carl G. 1945a. Studies in the logic of confirmation (I). *Mind* 54: 1–26. Reprinted, with some changes, in Hempel 1965, 3–51.

———. 1945b. Studies in the logic of confirmation (II). *Mind* 54: 97–121. Reprinted, with some changes, in Hempel 1965, 3–51.

———. 1960. Inductive inconsistencies. *Synthese* 12: 439–469. Reprinted, with some changes, in Hempel 1965, 53–79.

———. 1965. *Aspects of scientific explanation and other essays in the philosophy of science.* New York: The Free Press.

Hon, Giora. 2003. The idols of experiment: Transcending the "etc. list". In *The philosophy of scientific experimentation*, ed. Hans Radder, 174–197. Pittsburgh: University of Pittsburgh Press.

Howson, Colin, and Peter Urbach. 2006. *Scientific reasoning. The Bayesian approach.* 3rd ed. Chicago: Open Court.

Mayo, Deborah G. 1996. *Error and the growth of experimental knowledge.* Chicago: University of Chicago Press.

Popper, Karl. 1992. *The logic of scientific discovery.* New York: Routledge.

Toulmin, Stephen E. 1958. *The uses of argument.* Cambridge: Cambridge University Press.

Chapter 3
Theories of Confirmation in Which Hypotheses Have Probabilities and Inference to the Best Explanation

Abstract In this chapter, I discuss Bayesianism and Peter Achinstein's theory together since in both theories hypotheses have probabilities (i.e., p(h) = r; $0 \leq r \leq 1$). I also consider Inference to the Best Explanation as a method of theory choice since Achinstein has incorporated the requirement of an *explanatory connection* between hypothesis and evidence into his theory. I attempt to provide sufficient background to argue in later chapters that they do not provide satisfactory theories of evidence for clinical medical science. I argue that the concepts of explanation employed by Achinstein and Inference to the Best Explanation are different, and that at least the explanatory concept in the latter is not necessary for confirmation. After presenting part of a theory of explanation advanced by Bas van Fraassen, I argue that correct hypotheses are sought in clinical medical science, and that explanations are secondary.

3.1 General

Bayesianism and certain types of evidence in Peter Achinstein's theory of evidence require that for *e* to be evidence that *h*, *e* must raise the probability of *h*. In addition, Achinstein adds an "explanatory connection" requirement for *e* to be evidence that *h* in those types of evidence. In *Inference to the Best Explanation*, explanatory considerations predominate in hypothesis or theory choice.

3.2 Bayesianism

The Bayesian theory of confirmation rests on the notion that a confirmatory piece of evidence *e* raises the probability of a hypothesis *h*. Conversely, *e* could disconfirm *h* if it lowers *h*'s probability. Bayesianism derives its name from Bayes' theorem, which is derivable from the axioms of probability theory. Bayes' theorem can be written thusly:

© Springer Nature Switzerland AG 2020
J. A. Pinkston, *Evidence and Hypothesis in Clinical Medical Science*, Synthese Library 426, https://doi.org/10.1007/978-3-030-44270-5_3

$$P(h/e) = P(e/h) P(h)/P(e) \qquad (3.1)$$

where P(h/e) is read as "the probability of h given (or, conditioned on) e". Similarly, P(e/h), P(h), and P(e) are read as "the probability of e given h", "the probability of h", and "the probability of e", respectively. P(h) and P(e) are *unconditional* probabilities. P(e/h) is technically defined as the *likelihood* of h, and is the probability that h confers on e (Sober 2008, 9–10). It has also been referred to as the likelihood of e on h (Earman 1992, 34).

In the Bayesian approach, a hypothesis h has a *prior probability*, P(h), prior to the acquisition of evidence e. After e is acquired, we have the *posterior probability* P (h/e), the probability of h given e. In the simple case, P(h/e) can become a new "prior probability", on which some further e could act to affect its probability. The Bayesian definitions of confirmation are as follows (Howson and Urbach 2006, 91–92):

e *confirms or supports* h just in case $P(h/e) > P(h)$

e *disconfirms* h just in case $P(h/e) < P(h)$

e *is neutral towards* h just in case $P(h/e) = P(h)$

Bayes' theorem as written above relates four quantities *synchronically*, so that if three quantities are known, the fourth can be calculated. Bayesianism involves a rule for updating, and describes how probabilities should be related *diachronically* (Sober 2008, 11).

To illustrate, assume that a physician is talking to a patient about the results of his tuberculosis (TB) test, which is a chest radiograph (chest x-ray). Further assume that the ability of a chest x-ray to correctly identify a person with TB in the population has been determined empirically (Brown and Hollander 1977, 26–28; Yerushalmy et al. 1950, 453):

In Table 3.1, let h = person has TB, ~h = person does not have TB, e = positive x-ray reading, and ~e = negative x-ray reading. Assuming no other information or evidence, the probability of h, that the patient has TB, is the prevalence of TB in the population, 30/1820, or P(h) = .016. If the patient tests positive, the probability of that event occurring, the evidence, is 73/1820, or P(e) = .040. Since the test can be falsely positive or falsely negative, the unconditional P(e) is the average probability

Table 3.1 False-positive and false-negative errors in a study of X-ray readings

	Persons without TB	Persons with TB	
Negative x-ray reading	1739	8	1747
Positive x-ray reading	51	22	73
	1790	30	1820

Sources: B.W. Brown, Jr., and M. Hollander. Statistics: A Biomedical Introduction. New York: Wiley, 1977. p. 26, table 6. © John Wiley & Sons Ltd. Reproduced with permission of the Licensor through PLSclear. Also J. Yerushalmy et al. *The Role of Dual Reading in Mass Radiography*. Amer Rev Tuberc 61 (1950). p. 453, table 6

of obtaining e under the two alternative hypotheses, with TB (h) or without TB (~h):
P(h/e) = P(e/h) P(h) / [P(h) P(e/h) + P(e/~h) P(~h)] (Sober 2008, 13–15).

Substituting, we have : $P(h/e) = (.733) (.016)/[(.733) (.016) + (.028) (.984)] = .30$.

Based on acquisition of e, a Bayesian agent would update his prior probability, .016, to his posterior probability, .30. Thus, in this case, e confirms h.

If instead the patient had tested negative, then the probability of having TB can be determined:

$$P(h/{\sim}e) = P({\sim}e/h) P(h)/[P({\sim}e/h) P(h) + P({\sim}e/ {\sim}h) P({\sim}h)]$$

$$P(h/{\sim}e) = (.267) (.016)/[(.267) (.016) + (.972) (.984)]$$

$$P(h/{\sim}e) = .004$$

Since our posterior probability, P(h/~e), equals .004, and is less than our prior probability of .016, a negative x-ray reading (~e) disconfirms h.

In the above example, both a subjective and objective interpretation can be given. On a subjective view, the degree of belief in h has been increased by the acquisition of e, but there is also an objective increase in probability based on actual frequency data.

Bayesians use the laws of probability as coherence constraints on rational degrees of belief in their inductive logic of confirmation, and one justification for this that has been advanced are *Dutch Book arguments*, which relate degrees of belief with willingness to wager. These arguments are designed to show that if the laws of probability are not followed, losses will inevitably occur. For example, let us assume that a person with degree of belief p in sentence S is willing to pay up to and including $p for a unit wager on S (i.e., a wager that pays $1 if S is true), and is also willing to sell such a wager if the price is equal to or greater than p. A *Dutch Book* is a combination of wagers that, on the basis of deductive logic alone, can be shown to entail a sure loss. Talbott (2015) provides an example:

> Suppose that agent A's degrees of belief in S and ~S (written db(S) and db(~S)) are each .51, and, thus that their sum 1.02 (greater than one). On the behavioral interpretation of degrees of belief..., A would be willing to pay db(S) × $1 for a unit wager on S and db(~S) × $1 for a unit wager on ~S. If a bookie B sells both wagers to A for a total of $1.02, the combination would be a synchronic Dutch Book – synchronic because the wagers could both be entered into at the same time, and a Dutch Book because A would have paid $1.02 on a combination of wagers guaranteed to pay exactly $1. Thus, A would have a guaranteed net loss of $.02

Several virtues have been ascribed to Bayesianism. One source of support for the theory is that it can explain the role of hypothetico-deductive explanation in confirmation. When h logically entails e, e confirms h. This obtains because e has ruled out ~e, which would reduce the probability of h to zero.[1] In Bayesian terms, ~e reduces the probability of h to zero, providing maximum disconfirmation (Howson and

---

[1]Howson and Urbach (2006, 94) explain that when h logically entails e, p(e/h) = 1; thus if 0 < p (e) < 1, and p(h) > 0, then p(h/e) > p(h), which means that e confirms h.

Urbach 2006, 93–94). Nevertheless, as noted earlier, h is usually conjoined with auxiliary hypotheses a: (h&a) → e, and disconfirmation might actually be due to a false a, and leave h unscathed (Earman 1992, 63–65).

Another virtue claimed for Bayesianism is that it may solve the riddle of the Raven Paradox, since Bayesian confirmation is a matter of degree. If one accepts that (1) hypotheses of the form "All Rs are Bs" is confirmed by the finding of an R that is also a B, and, (2) logically equivalent hypotheses are confirmed by the same evidence, then one may be led to the conclusion that an object that is both non-black and non-raven is confirmatory of the hypothesis "All ravens are black" as would be the finding of a black raven. That, for example, finding a white shoe is also confirmatory of the hypothesis. Bayesians concede that both are confirmatory, but finding a black raven carries more force and is confirmatory to a much greater degree, since the class of non-black things is vastly more numerous than the class of ravens. So if we are sampling from the class of ravens rather than the class of non-black things, we are more likely to produce a non-black raven, and the greater confirmatory power derives from the relative threat of falsification (Suppes 1966).[2]

Bayesianism is said to illuminate the problem of irrelevant conjunction. Consider that if hypothesis h logically implies evidence e, then any conjunct i when conjoined with h also implies e, even a totally irrelevant conjunct. Thus, e confirms both h & i and h. But Bayesian incremental confirmation is proportional to the prior probabilities of h and h&i; in general, $P(h\&i) < P(h)$, so adding the irrelevant conjunct i to h lowers the incremental confirmation afforded by e (Earman 1992, 64–65).

Bayesianism has also been said to address the old and widely held idea that different and varied evidence supports a hypothesis more than a similar amount of homogeneous evidence. Thus, if two items of evidence, e_1 and e_2, are similar, then $P(e_2/e_1)$ may approximate 1. For example, suppose we have the observation that a pebble fell to the ground from a certain height in two seconds today, and is similar to the pebble's fall yesterday; here e_2 provides little additional support when e_1 is known. But e_1 and e_2 are both different from, say, the earth's trajectory around the sun or the effect of the moon on the tides. Thus when the pieces of evidence are dissimilar, then $P(e_2/e_1)$ may be significantly less than 1, so that e_2 adds a greater degree of confirmation to that already supplied by e_1. This allows evidence to be analyzed in terms of degree of difference (Howson and Urbach 2006, 125–126).

Bayesianism has also been criticized on several grounds as being inadequate as a theory of scientific confirmation. One criticism is that it is too weak; all that is required for confirmation of theory T by a piece of evidence is an increase in T's probability over its prior probability. Thus, my buying one ticket in a lottery where the odds of winning are ten million to one is evidence that I will win the lottery, thus confirming that hypothesis. Granted it is not much evidence, but it is *some*. Likewise, for example, assuming some finite probability that a person who swims in the ocean will be attacked by a shark, my swimming in the ocean confirms the hypothesis that I

---

[2]The extent to which Bayesianism may offer a solution to the Raven Paradox is controversial. For further discussion see, e.g., Earman (1992, 69–73) and Howson and Urbach (2006, 99–103).

will be attacked by a shark. This notion could be generalized to virtually any activity carrying any risk whatever, like being a passenger in an airliner. Thus, it is claimed that any notion of evidence or theory of confirmation that only requires an increase in T's probability is too weak to be taken seriously (Achinstein 2001, 6).

Reasons have also been advanced for why the use of betting odds or Dutch Book arguments are far from conclusive as an accurate measure of the degrees of belief of a rational agent. There are many cases of propositions in which we may have degrees of belief for which no wager will be offered; and, we may have values other than the values we place on gambling odds and these may affect our decision even to gamble. Also, we could avoid a possible loss by refusing to gamble at all, even if the odds were in our favor (Glymour 1980, 71–72). In addition, it has been argued that betting behavior may only be indicative of, and not constitutive of, underlying belief states, and that actual betting behavior by gamblers and laymen is often at variance with that predicted by Dutch Book arguments. In poker, for example, betting high may be a good way to scare off other players and win the pot (Earman 1992, 40–43). If betting behavior actually was an accurate measure of rational belief, it would be difficult to explain why millions of people bet $1 on a lottery ticket when the chance of winning is, say, one in 16 million.

Bayesian principles have been advanced as providing a unified scientific method (Howson and Urbach 2006, 91), but there is a substantial difference between the relatively straightforward case of TB diagnosis illustrated above and the use of Bayes' theorem in testing a deep and general scientific theory like Darwin's theory of evolution or Einstein's GTR. To illustrate, in the case of the GTR there are no frequency data on which to rely, and a problem arises in attempting to quantify the probability of the evidence in the case of, for example, Eddington's data on the bending of starlight during a solar eclipse. Although a value could be assigned to P(e/GTR), how is *not*-GTR ($\sim$GTR) to be evaluated? It consists of all the theories (T_1, $T_2, \ldots, T_n$) that are incompatible with the GTR (and when taken together, is $\sim$GTR, and is called the *catchall* hypothesis), some of which presumably have not even been formulated yet. The likelihood of $\sim$GTR is the average likelihood of these specific alternatives, weighted by the probability they have, conditioned on the GTR being false:

$$P(\text{observation}/\sim \text{GTR}) = \Sigma_i P(\text{observation}/T_i)\, P(T_i/\sim \text{GTR})$$

How can this be objectively quantified? Although subjective Bayesians may give likelihoods reflecting personal degrees of confidence, it would seem that we should expect more (Sober 2008, 28–29; Earman 1992, 117).

3.3   Achinstein's Theory of Evidence

Peter Achinstein (2001) presents and elaborates in detail his theory of evidence in which for a fact *e* to be evidence for a hypothesis *h*, *e* must provide *a good reason to believe h*. Thus, he rejects Bayesianism, hypothetico-deductivism, and Hempel's satisfaction theory as being too weak to accurately reflect scientific practice and the way that most scientists assess evidence in support of hypotheses. Hypothetico-deductivism fails because e, if true, is evidence that h if and only if h entails e, which would, for example, allow that since the fact that light travels in a straight line is derivable from the classical wave theory of light, it is evidence that light is a classical wave motion. Hempel's theory allows that the observation of a black raven is evidence for the hypothesis that "All ravens are black". In neither case, Achinstein (2001, 6) contends, does the evidence provide a *good reason to believe* the hypothesis.

Achinstein postulates that scientists use both objective and subjective concepts of evidence, but that the most important concept for them is objective. He outlines his concept of subjective evidence as being relativized to a specific person or group, and also may be evidence for that person or group at one time but not another. Subjective evidence satisfies at least three conditions, (2001, 23):

1. X believes that e is evidence that h;
2. X believes that h is true or probable; and
3. X's reason for believing that h is true or probable is that e is true.

In the above, e does need not be true, but only that X believes it is. Evidence e is accepted until new evidence refutes it. It requires belief, and that someone or some group is in a certain epistemic situation in regard to the evidence.

An example of subjective evidence that he provides is that of Heinrich Hertz's evidence for the neutrality of cathode rays in 1883 that was based on a flawed experimental setup. Hertz concluded (wrongly) at the time that cathode rays were not electrically charged, but later in 1897 J. J. Thomson repeated the experiment under conditions satisfactory for demonstrating that cathode rays were indeed negatively charged. Hertz's experimental results were subjective evidence for the neutrality of cathode rays from 1883 to 1897, but not thereafter (Achinstein 2001, 13–24).

Achinstein develops three objective concepts of evidence that he believes scientists also employ: *epistemic situational* (ES), *veridical*, and *potential*. ES-evidence is of the sort obtained by Hertz in 1883, but it is relativized to a *type* of epistemic situation, and there is no requirement that anyone *be* in that situation. Evidence e can be ES-evidence for h even if no one in fact believes that e or h is true or believes that e is ES-evidence for h. Unlike subjective evidence, ES-evidence requires that e be true, and that the person or group whose evidence it is be justified in believing h on the basis of e (2001, 19–22).

Veridical evidence requires that h be true, and that e provides *a good reason to believe h*. Evidence later rejected as wrong based on new findings is not evidence and never was. "Not a good reason to believe" is Achinstein's way to describe this, unlike "justification" in the subjective and ES cases. Thus "good reason to believe" functions like a "sign" or "symptom". For example, a rash may be a good reason to believe that a certain disease is present, even if no one is aware of the connection. The epistemic situation that pertains at any time is irrelevant. Veridical evidence can also depend on empirical facts not reported in e or h. Veridical evidence is distinguished from *conclusive* evidence: conclusive evidence establishes the certain truth of h, i.e., $P(h/e) = 1$. If e is veridical evidence with respect to h, then h is true, but it does not have to establish that h is true with certainty (2001, 24–27).

Potential evidence is like veridical evidence, but weaker. It does not require h to be true, only *probably* true. It is *fallibilist*. It is not relativized to an epistemic situation, as in ES-evidence, but, like ES-evidence, it requires e to be true (2001, 27–28).

Achinstein advances a new interpretation of probability that he calls *objective epistemic probability*. It attempts to deal with the question of how reasonable it is to believe a proposition: what does it mean to say that it is reasonable to believe something? It can be relativized to other beliefs held by someone, or, say, some group. It can also be abstract, and not dependent on other beliefs, for example a belief about the number of heads in a coin tossing experiment. It admits of degrees, hence is subject to the rules of probability. Achinstein's concept of probability is not a measure of *how much* belief one has or ought to have, or *how strong* the belief is or ought to be, but rather it is *how reasonable it is to believe something* that is subject to differences of degree. There is no relativization to persons, so it differs from a subjective view of probability. It differs from Carnap's view that beliefs come in degrees; e.g., for Carnap, if $P(h) = r$, then one is justified in believing h to the degree r. It also differs from frequency and propensity views of probability, which are objective views about the world, and not about reasonableness of belief. For Achinstein, a probability statement of the form "$P(h) = r$" is understood to mean that: The degree of reasonableness of believing h is r. Also, if the probability of $h = \frac{3}{4}$, then it is three times more reasonable to believe that h is true than that h is false (2001, 95–100).

For Achinstein, although evidence is related to probability, an increase in probability is neither necessary nor sufficient. On his view (2005, 44),

(a) For a hypothesis h and putative evidence e, if e is a good reason to believe h, then there is some number k (greater than or equal to 0) such that p(h/e) > k.
(b) If e is evidence that h, then e must be a good reason to believe h.
(c) If e is a good reason to believe h, then e cannot be a good reason to believe not-h. He states that assumptions (a)–(c) can be shown to require that k in (a) be ½, so that
(d) e is evidence that h only if p(h/e) > ½.[3]

In addition to raising the probability of h to above ½, there needs to be some *relevance* relation, or some *connection* between h and e. Consider:

e: Michael Jordan eats Wheaties
b: Michael Jordan is a male basketball star
h: Michael Jordan will not become pregnant

---

[3] As Achinstein notes, whether (a), (b), and (c) require (d) depends on his argument in his *Book of Evidence* (2001, 115–116).

Here, e, b, and h are the evidence, background information, and hypothesis, respectively. Evidence e provides no additional reason to believe h, since b alone makes h's probability approximately one, since it asserts that Jordan is male (2001, 145–146). It is irrelevant; it provides no explanatory connection between h and e. Thus he develops the notion of an *explanatory connection* between h and e, which he defines by reference to a *correct explanation*:

> There is an explanatory connection between h and e if and only if either h correctly explains why e is true, or e correctly explains why h is true, or some true hypothesis correctly explains why h is true and why e is true (2001, 160).

His notion of "correct explanation" has three characteristics:

(1) It is objective: whether h correctly explains e does not depend on what anyone knows or believes
(2) It is non-contextual: in this sense, it is like causation, in that it does not vary with the interests and knowledge of different inquirers
(3) It does not depend on any requirement of evidence, nor invoke other explanatory or evidential concepts

Achinstein wishes to avoid some evidence e which could appear, for example, as an irrelevant conjunct: In the above Michael Jordan example, p(h/e&b) = p(h/b), which is approximately one, and p(h/e) = p(h). Unlike causation, he asserts that probability is not *selective*: "If some fact or event E_1 caused an event E, it will be false to say that E_1 and E_2 caused E, even if E_2 is a fact or is an event that occurred, unless E_2 was causally involved in producing E" (146). But for probability, an irrelevant conjunct could appear and the probability of h be unaffected: p(h/e&b) = p(h/b). To avoid this, he requires an explanatory connection between h and e.

Achinstein's necessary and sufficient conditions for potential evidence are (2001, 170):
(PE) e is potential evidence that h, given b, only if

(1) p(there is an explanatory connection between h and e/e&b) > ½
(2) e and b are true
(3) e does not entail h.

His definitions of veridical, ES, and subjective evidence are (2001, 174):

e is veridical evidence that h, given b, if and only if
1. e is potential evidence that h, given b
2. h is true
(3. There is an explanatory connection between e's being true and h's being true.)

e is ES-evidence that h (with respect to an epistemic situation ES) if and only if e is true and anyone in ES is justified in believing that e is (probably) veridical evidence that h.

e is X's subjective evidence that *h* at time *t* if and only if at *t*, X believes that *e* is (probably) veridical evidence that *h*, and X's reason for believing *h* is true (or probable) is that *e* is true.

Does Achinstein's explanatory connection condition somehow establish a necessary role for explanation before we can have a good reason to believe a hypothesis? Consider a hypothesis H that might be encountered in clinical medical science:

H: X will experience outcome Y from intervention Z

Here, X represents an individual patient, Z is an intervention, and Y is the expected outcome experienced by patient X due to (because of) intervention Z.

Although H is a perfectly good therapeutic hypothesis, in modern clinical medical science hypotheses such as H are treated as predictions based on evidence, for example, studies such as RCTs, which have established the extent to which we can expect outcome Y from intervention Z in an individual patient.

Achinstein conveniently provides an example of H from clinical medical science in discussing his explanatory connection (2001, 156):

John, who has symptoms S, presumably is taking medicine M in hopes of getting relief from symptoms S. Background information *b* includes the facts that 80% of patients with symptoms S that take M get relief in a week (R), so $p(R/S\&M) = .8$, and also that among those with S who take M and get relief in a week, 70% do so *because* they took M, so $p(R$ because of $M/R\&S\&M) = .7$. He then shows that p (R because of $M/S\&M) = .56$. Thus, letting

e: John, who has symptoms S, is taking M
h: John will get relief in a week
b: the probability information given above

Achinstein concludes from this example that $p(h$ because of $e/e\&b) = .56$, which meets his criteria for "a good reason to believe" *h* (although he says that it is not particularly strong evidence, since only a little more than half of those with symptoms S who take M get relief because of M). To have a good reason to believe *h* does not only consist in *e*'s raising the probability of *h* to greater than ½, but also the probability that the *reason* that John's symptoms will be relieved by M, given *e&b*, is greater than ½.

The explanatory connection required by Achinstein seems established in this example, since it is based on the probability information provided in *b*, in which the putatively correct explanation is objective, noncontextual, and not explicated in terms of evidence or other explanatory concepts. But the important point for our discussion is to note that the basis for inferring or establishing an explanatory connection between *h* and *e* comes from background information *b*. But how would we acquire the kind of information provided in *b* in this example? That is, that 80% of patients with symptoms S taking medicine M get relief in a week, and that 70% of those who get relief do so *because* they took M?

It obviously could come from studying a group or groups of patients, either from observation or from one or more clinical trials. Since *b* is assumed by Achinstein to be true, presumably he would not object to our assuming that it came from an RCT, say, like the following:

An RCT was conducted to test the efficacy of medicine M in producing relief for patients with symptoms S. A total of 200 patients with symptoms S were randomized to receive M or a placebo. At the end of one week, among 100 patients randomized to M, 80% got relief. Among 100 controls, the comparable figure was 24%. Conclusion: M was 56% efficacious in relieving symptoms S within a week. Another 24% got relief in a week for other reasons.

It seems from the above that the *explanatory* element in the explanatory connection between *h* and *e* is derived from the results of one or more studies that have tested the therapeutic hypothesis regarding the efficacy of M in patients with S. This is the evidentiary basis for believing *h*: John will get relief in a week. The explanatory connection between *h* and *e* does exist, apparently, because *e* correctly explains (or might) why *h* is true. But the correct explanation, hence the explanatory connection, is on the previously confirmed hypothesis concerning the efficacy of M in patients with S. It is *derivative*: it depends on confirmation. Another of Achinstein's examples can also be used to illustrate this (2001, 152):

e: Arthur has a rash on his arm that itches.
b: Arthur was weeding yesterday bare-armed in an area filled with poison ivy, to which he is allergic.
h: Arthur's arm was in contact with poison ivy.

In this example, Achinstein contends that given *e* and *b*, the probability is high that the reason (*e*) Arthur has a rash on his arm that itches is that (*h*) his arm was in contact with poison ivy. Thus, under these conditions, given *e* and *b*, *e* is a good reason to believe *h*. That is, p(*h/e*) is high, and p(there is an explanatory connection between *h* and *e/e&b*) is high. But why is this so? Because of a previously highly confirmed hypothesis, which is assumed but is unstated, that I will call h_1:

h_1: There is a high probability that persons allergic to poison ivy will develop a rash that itches on their skin if it is exposed to poison ivy

Here, h_1 is assumed as part of *b*. Without h_1, *b* loses its force in providing the needed explanatory connection between *e* and *h*. Thus, it is *derivative*: it depends on confirmation.

It appears that the *explanatory* element in the explanatory connection between *h* and *e* is based on one or more confirmed hypotheses, which establishes in these cases his desired relevance relation between *h* and *e*. Achinstein's notion of explanation as used here is objective and non-contextual, and does not depend on what anyone knows or believes. Thus, it is a different concept of explanation than that employed in Inference to the Best Explanation, which is context-dependent and does depend on the knowledge and interests of different inquirers. I will argue that at least this latter notion is not necessary for confirmation in clinical medical science.

3.4   Inference to the Best Explanation

Gilbert Harman argued that various forms of non-deductive inference, including "abduction," "enumerative induction," and "eliminative induction," correspond approximately to what he calls *The Inference to the Best Explanation*. To explain certain evidence, one forms a hypothesis, and makes an inference to the truth of that hypothesis. In general, more than one hypothesis could explain this evidence, and therefore one must first eliminate rival hypotheses so that the "best" hypothesis is chosen. Harman notes that arriving at the best explanation presumably involves considerations such as simplicity, plausibility, being less *ad hoc*, explaining more, and so forth, but concedes that the actual process is problematic and not well understood (1965, 89).

Inference to the Best Explanation as a description of our inductive practices is attractive because it plausibly seems to account for how we go about making everyday inferences. When a detective puts together all of the evidence in a certain case and infers that a particular suspect *must* be guilty, he is reasoning that no other explanation is sufficiently simple or plausible to be accepted. When a physician examines a patient with a characteristic rash and diagnoses measles, he has essentially eliminated rival hypotheses and inferred that measles is the best explanation for the evidence before him. Beginning with the evidence, we infer, what would, if true, provide the best explanation of that evidence.

In the case of causal explanation, Harman argues, a better account of inference emerges if "cause" is replaced by "because." We infer not only statements of the form X *causes* Y but, more generally, statements of the form Y *because* X or X *explains* Y. Here, inductive inference is construed as inference to the best of competing explanatory statements, of which inference to a causal explanation is a special case. Thus, one might infer that a certain mental state explains someone's behavior, but such an explanation by reasons might not be causal explanation (1973, 130).

Peter Lipton (1991) has further explored Inference to the Best Explanation. He notes that the attempt to justify Inference to the Best Explanation, which itself is an inductive process, runs up against Hume's objections, and that any such attempt is fraught with the problem that it is itself inductive. Unlike deduction, we are faced with underdetermination: even if our premises are true, we can have a false conclusion (1991, 6–8). Thus Lipton focuses on trying to describe the process of Inference to the Best Explanation, and avers that explanatory considerations are our guide to inference. He argues that in many ways, explanatory considerations guide inquiry into determining what inferences to make. They tell us what to look for and whether we have found it. The physician *infers* that the patient has measles because this is the best explanation for the evidence before him (1991, 57).

Lipton stresses that the best explanation is selected from among the pool of *potential* explanations (1991, 59). Inference to the Best Explanation cannot be understood as inference to the best of the *actual* explanations, since that would make us too good at inference and make our inferences true. We also should wish to consider only *plausible* explanations and not just *possible* explanations, since many possible explanations could be wildly off the mark. Lipton considers two notions of "best": *likeliest* and *loveliest*. The likeliest is the most warranted; the loveliest is the one that, if true, would be the most explanatory or provide the most understanding. Aspects of loveliest include broadly aesthetic considerations like theoretical elegance, simplicity, and unification. Likeliness is relative to the total available information, whereas loveliness may not be. If we choose likeliest as the best, it may be trivial since we still need more than the likeliest *cause*. We should show how likeliness is determined at least in part by explanatory considerations. So, we should choose *loveliest*. Scientists also entertain these broadly aesthetic considerations (1991, 61–64).

To explain why P rather than Q, we need to find a causal difference between P and not-Q ($\sim$ Q), consisting of a cause of P and the absence of a corresponding event in the history of $\sim$ Q. Therefore we explain why Jones rather than Smith got paresis, since only Jones had syphilis. Of course our interests determine the foil that we use in contrastive inference, and our explanations may be different when different foils are used. Thus, our explanation of why Jones got paresis and Smith did not because Jones had syphilis and Smith did not, will not suffice to explain why Jones got paresis and Smith did not in the event that *both* Jones and Smith had syphilis (1991, 43–44). Thus, it seems that for Lipton, explanation is highly *context dependent*.

To illustrate his notion of contrastive inference, Lipton uses the example of the investigation of childbed fever by the Austrian physician Ignaz Semmelweiss (Hempel 1966, 3–8). Between 1844 and 1848, while working in a Viennese hospital, Semmelweiss observed that the women in the First Maternity Division got the disease at a much higher rate than women in the Second Division. He developed three types of hypotheses: (1) Hypotheses that did not mark a difference between the divisions, and so were rejected (e.g., "epidemic influences" descending over the entire area); (2) Hypotheses that did mark a difference between the divisions, but where eliminating the difference in putative cause did not affect the difference in mortality. For example, women in the First Division were delivered lying on their backs, while women in the Second Division were delivered lying on their sides; but when Semmelweiss arranged for all women to be delivered on their sides, there was no change in the disease rates; and, (3) Hypotheses that marked a difference between the two groups, and where eliminating the difference also eliminated the difference in disease rates. One of Semmelweiss's colleagues got a puncture wound after doing an autopsy, and died of an illness with symptoms similar to those of childbed fever. Semmelweiss hypothesized that "cadaveric matter" was the etiology. Medical students did the deliveries in the First Division after doing autopsies. However, midwives did the deliveries in the Second Division, and the midwives did not do autopsies. So he had the medical students disinfect their hands before deliveries, and the excessive number of childbed fever cases in the First Division went down.

Although Hempel used the Semmelweiss example as a paradigm of the hypothetico-deductive method, Lipton suggests that Inference to the Best Explanation works better. It illuminates the context of discovery and how hypotheses are generated, something that hypothetico-deductivism rejects. Explanatory considerations focus and direct inquiry. Potential hypotheses are not assembled and deductively rejected. Lipton contends that Popper is wrong that disconfirmation works through refutation. Scientists reject theories as false not because the evidence refutes them, but because they fail to explain the salient contrasts. Thus, for example, the hypothesis of "epidemic influences" was not rejected because the evidence contradicted it. Like any epidemic, Lipton argues, some people get sick and others do not. So the hypothesis does not entail that the mortality in the two divisions is the same. Semmelweiss rejected hypotheses because they failed to explain contrasts, not because they were logically incompatible with them (1991, 88–94).

Bas van Fraassen has criticized Inference to the Best Explanation on several grounds. One criticism has been called the "bad lot" argument (van Fraassen 1989, 142–143). Inference to the best Explanation selects one hypothesis among those that are available to us. So, it is possible that our selection may be nothing more than the best of a bad lot. Although selecting the "best" hypothesis from among a set of rivals is a "weighing" of the evidence (and, justifiably so) within the set, the inference to the best explanation requires another step – an ampliative step. To infer that some hypothesis is more likely to be true than not requires the prior belief that the truth is already more likely to be found in the set than not. This cannot be justified based on some notion of a "privilege of our genius" at being naturally endowed, with faculties that lead us to hit on the right range of hypotheses. Nor can it be justified that we *must* select the best on the basis of some rule of right reason (van Fraassen 1989, 143–145). In this connection, Psillos (1996, 37) has argued that van Fraassen has ignored background information that scientists use to arrive at a set of plausible potential hypotheses, which he calls the *background knowledge privilege*.

Another argument against Inference to the Best Explanation is the *argument from indifference*. Van Fraassen says, "... there are many theories, perhaps never yet formulated but in accordance with all evidence so far, which explain at least as well as the best we have now. Since these theories can disagree in so many ways about statements that go beyond our evidence to date, it is clear that most of them by far must be false. I know nothing about our best explanation, relative to its truth-value, except that it belongs to this class. So I must treat it as a random member of this class, most of which is false. Hence it must seem very improbable to me that it is true" (van Fraassen 1989, 146). Psillos (1996, 43) alleges that for this argument to have force, one must first show that there always are other potentially explanatory hypotheses waiting to be discovered, in addition to explaining the evidence at least as well. Ladyman et al. (1997, 309) point out, however, that even if it were the case that none of the unborn hypotheses offers a better explanation of the evidence than the best of the lot we now have, Inference to the Best Explanation would still be unacceptable. This is because it would require the additional premise that there is (almost) always a unique best explanation, that explanation would be ranked or ordered according to some standard of "goodness" for which there would exist a greatest element. But what justification do we have that this is so?

Van Fraassen does not deny that there is a common sense element in Inference to the Best Explanation that must be respected. He says, "If I already believe that the truth about something is likely to be found among certain alternatives, and if I want to choose one of them, then I shall certainly choose the one I consider the best. *That is a core of common sense which no one will deny*" (van Fraassen 1989, 149). He objects to the idea that Inference to the Best Explanation can be considered a rule to form warranted new beliefs on the basis of the evidence, and the evidence alone, in a purely objective manner. In Inference to the Best Explanation, hypotheses are evaluated by how well they explain the evidence, where explanation is an objective relation between hypothesis and evidence alone (1989, 142).

3.4.1 What Is an Explanation?

Several philosophical theories of explanation have been proposed (e.g., see Achinstein 1983), but no theory has as yet gained general acceptance. I will briefly discuss part of one such theory, the one proposed by van Fraassen (1980, 97–157), which will help lead us into a discussion of the role of explanation in clinical medical science.

For van Fraassen, an explanation is an answer to a "why" question. Thus, a theory of explanation must be a theory of why-questions (1980, 134). For example, consider the question: *Why is this conductor warped?* The question implicitly assumes that the conductor is warped, and the questioner is asking for a reason why. The proposition that the conductor is warped is the *topic* of the question. Next we have a *contrast class* for this question, which is a set of alternatives. For example, one contrast for this topic could be why is *this* conductor warped instead of that one; another contrast could be why is this conductor warped and not straight. Finally there is a *relevance relation*, which is the respect-in-which a reason is requested and what determines what is to count as a possible explanatory factor or set of factors. So in the case of the warped conductor, the request might be for the events leading up to the warping. In this case, relevant factors might include an account of human error, switches being closed, or moisture condensing in the switches, for example. Thus a why-question Q expressed by an interrogative in a given context will be determined by three factors: a *topic* P, a *contrast-class* X consisting of $\{P_1, \ldots, P_k, \ldots\}$ and a *relevance relation* R.

How would an answer to a why-question look given this schema? A direct answer, van Fraassen writes, would look like this (1980, 143):

P_k *in contrast to* (the rest of) X *because* A.

This sentence must express a proposition. What is claimed here is that P_k is true, that the other members of the contrast class are not true, that A is true, and that A is a *reason*.

Consider an example of a why-question Q. Q: Why does Smith have a swollen ankle (as opposed to his ankle's not being swollen)? A: (Because) he suffered an inversion sprain. Here the conditions are met:

P_k: Smith's ankle is swollen (true)

(the rest of) X: Smith's ankle is not swollen (untrue)

A: He suffered an inversion sprain (true)

And, A is a *reason*.

The relevance relation R could come into the schema perhaps as "events leading up to" P_k. Relevant here might be the following: Smith stepped off a curb, and as his foot was coming down his ankle inverted and bore the full weight of his body, producing a severe sprain.

As van Fraassen notes, we evaluate the answers to why-questions in the light of accepted background theory as well as background information (1980, 145). So the answer to Q may involve the accepted background theory of the pathophysiology of inversion sprains of the ankle: The force of the weight of the body coming down on the ankle causes tearing and disruption of the soft tissues of the ankle, which include blood vessels and lymphatic vessels; these events lead to bleeding and the accumulation of fluid under the skin surrounding the ankle, thereby producing the swelling.

Van Fraassen also considers how description and explanation relate to scientific theory. He maintains that whereas description is a two-term relation between theory and fact, explanation is a three-term relation between theory, fact, and context. Being an explanation, he says, is essentially relative, for an explanation is an *answer*, and what answer is given depends on context (1980, 156). The why-question is a request for *information*, and that information is essentially *descriptive*. If a scientist is asked to explain something, the answer is not different in kind, nor does it sound or look different, than the information given when a description is asked for. If an economist were asked to explain the rise in oil prices, for example, the answer would consist of descriptive information such as changes in oil producers, oil supplies, and oil consumption. Scientific explanations do not differ in form from "ordinary" explanations; in general, the information comes from science (1980, 155).

Thus for van Fraassen,

> ... scientific explanation is not (pure) science, but an application of science. It is a use of science to satisfy certain of our desires: and these desires are quite specific in a specific context, but they are always desires for descriptive information ... The exact content of the desire, and the evaluation of how well it is satisfied, varies from context to context ... [and] in each case, a success of explanation is a success of adequate and informative description (1980, 156–157).

3.4.2 Evidence, Hypothesis, and Explanation in Clinical Medical Science

For the most part, the type of explanation sought in clinical medical science is *causal* explanation. "Why does this patient have these symptoms?" is a request for the cause of the patient's symptoms. "Why does my head hurt?" is a request for the cause of the patient's headache.

Consider the following clinical scenario concerning a diagnostic hypothesis:

James arrives at his physician's office complaining of malaise and a sore throat. On examination, he is found to be febrile, and his throat is inflamed with a white exudate. The clinical diagnosis is acute pharyngitis. Based on the physician's background information, she considers two hypotheses as to the etiology of the pharyngitis:

H_1: A bacterial infection, such as with streptococcus, is the cause
H_2: A viral infection is the cause

Evidence is sought to confirm or disconfirm these hypotheses. Results from a blood test strongly favor a bacterial infection. A throat culture is obtained, which grows out streptococci. Thus the weight of the evidence strongly favors bacterial infection with streptococci as the etiology. Therefore,

Q: Why does my throat hurt?
A: (Because) you have streptococcal pharyngitis

Clinical medical scientists, like all scientists, want their hypotheses and theories to be true. And those theories and hypotheses that are selected for clinical decision making, such as those that form the basis for therapy or prevention programs, are increasingly being selected based on the weight of evidence supporting them. While no one can argue that explanatory considerations are not important, the thesis that I wish to defend is this:

T: *Confirmation is prior to explanation*

More specifically, for the clinical medical scientist, the "best" explanation will be based on the hypothesis that is best confirmed by the evidence.

Gilbert Harman (1965) has advanced *Inference to the Best Explanation* as a method of hypothesis choice, and it has been further explicated and defended by Peter Lipton (1991). The basic approach is to begin with the evidence, and select the hypothesis that, if true, provides the best explanation. Since more than one hypothesis may explain the evidence, we must reject alternative hypotheses so that we are ultimately left with the hypothesis that best explains the evidence. We infer from the best explanation to the truth of the hypothesis.

Much has been written about how one decides which of the competing hypotheses provides the "best" explanation. Keith Lehrer (1974, 165) has remarked on the "hopelessness" that any useful analysis of the concept that one explanation is better than another may be forthcoming. Thagard (1978, 79) advances the notions of *consilience*, *simplicity*, and *analogy* as providing useful criteria for theory choice. Nevertheless, one starts with the evidence, and infers to the truth of the hypothesis that best explains it.

However, what is the relationship between confirmation and explanation? Here we are concerned with the hypothesis that is best confirmed by the evidence, and the "best" explanation. Does one precede (in time) the other? Can they be arrived at concurrently? Does one in some way *depend* on the other?

It would not seem that they are *causally* related. The hypothesis that is best confirmed does not cause the best explanation to be the best. Nor would it seem that the best explanation would cause the best confirmed hypothesis to be the best confirmed. But consider Harman's Y *because* X in the non-causal sense: Is the best confirmed hypothesis best confirmed because it provides the best explanation? Or, is the converse true, that the best explanation is best because the hypothesis is best confirmed? Or can these two in some way be contemporaneous?

Consider a pediatrician that has just been told by the nurse that the child in the next examination room has a rash. On entering the examination room, the pediatrician notes the characteristic rash of measles. Immediately, it seems, the pediatrician has both confirmed the diagnosis of measles and, potentially at least, provided at least one explanation for the rash. But an explanation arguably is, at least in part, an answer to a why-question. Must the question be explicitly asked? It might be plausible to suppose that the pediatrician implicitly desires a correct explanation for the rash, in addition to (or even in place of) making the correct diagnosis. Can we assume that the pediatrician is perhaps implicitly or unconsciously seeking an explanation?

It would seem more plausible to assume that the pediatrician primarily wishes to make the correct diagnosis. To the extent that explanation is desired or requested (e.g., "Why does my daughter have this rash?"), no doubt one candidate for the "best" explanation would be a correct one, commensurate with the diagnosis (e.g., "She has measles.").

This measles example seems closely related to Lipton's example of the snowshoe tracks (1991, 26). He imagines himself looking at snowshoe tracks in front of his house and being asked why they are there. He explains that someone wearing snowshoes has recently passed by, and asserts that this is a perfectly good explanation, even though he did not actually see the person making the tracks. Lipton goes on to say that these "self-evidencing" explanations are ubiquitous and may be perfectly acceptable, and that the circularity is benign.

Let us alter Lipton's example. Suppose I am alone and walk outside and see the snowshoe tracks. I simply know that someone has passed by wearing snowshoes. And suppose that I look up into the sky, and far up without hearing a sound, I see a characteristic condensation trail. I simply know that an airplane has passed overhead. Is there a role here, in these quotidian examples, for explanation or confirmation? My hypotheses are: "Someone has passed by wearing snowshoes" and "An airplane has passed overhead." My evidences are: "There are snowshoe tracks" and "There is a condensation trail overhead." My explanations for the evidences are: "Someone has passed by wearing snowshoes" and "An airplane has passed overhead." But notice that the statements expressing the hypotheses and those expressing the explanations are the same. For someone with the requisite knowledge, hypothesis and explanation may be one and the same. Hypothesis, evidence, and (at least, potential) explanation are all there contemporaneously. In the example from Lipton above, he is being asked for an explanation: someone has seen peculiar tracks, and apparently not knowing they are snowshoe tracks, asks for an explanation (e.g., why they are there). Similarly, for the pediatrician, having the requisite knowledge, hypothesis, evidence, and explanation are all there contemporaneously. However, the mother may ask, "Why does my daughter have this rash?" This, of course, is a request for an explanation.

These "self-evidencing" cases of contemporaneous hypothesis, evidence, and explanation such as our measles example are the exception in clinical medical science, and as a rule most diagnoses are not arrived at quite so easily. We may start with some evidence such as information provided by the patient along with various signs or symptoms, but most often, particularly for serious conditions, additional diagnostic testing must be done in the process of confirming or disconfirming various diagnostic hypotheses, such as in the above case of the patient with the sore throat. In that case two hypotheses were considered initially. The weight of the evidence favored the diagnosis of streptococcal pharyngitis, and that became the working diagnosis on which treatment decisions would be based. Is there some role here for explanation?

Maybe it would be possible to view the case of James as an exercise in explanation: James went to his physician seeking an explanation for his sore throat. The physician sought to explain his sore throat, and on examination her explanation for the soreness was acute pharyngitis. She considered two explanations, and the best explanation based on the evidence of the blood test results and throat culture was streptococcal pharyngitis. Thus she inferred the truth of the hypothesis of streptococcal pharyngitis because it was the best explanation of the evidence. Is this not a clear-cut case of Inference to the Best Explanation?

The more plausible interpretation, I would argue, is that James visited his physician primarily because he wanted treatment for his sore throat, in addition to wanting to know why it hurt. The physician wanted to make the correct diagnosis and institute appropriate treatment. We should note here however, that in this example that the best confirmed hypothesis is streptococcal pharyngitis: "The patient has streptococcal pharyngitis." If James were a bacteriologist, a candidate for the best explanation, should he ask why his throat hurts, might be "You have streptococcal pharyngitis." If James were a young adult instead, he might be satisfied with the relatively well-known lay term for this condition, "Strep Throat." If James were an inquisitive nine year-old, considerably more in the way of explanation may be required. In view of this dependence on context for explanation, for Inference to the Best Explanation as a theory of confirmation to work in this example, and bearing in mind that determining the "best" explanation is a central problem for this theory, it must be the diagnosing physician who must decide what is the best explanation, and as I said earlier, this will almost invariably be the one that is believed to be the correct one. And why, exactly, should we assume that an explanation is being sought rather than seeking the information necessary to make the correct diagnosis?

The primary aim in the diagnostic process is to make the correct diagnosis. The primacy of making the correct diagnosis derives from its overarching importance: It forms the basis for instituting treatment and providing prognostic information. It is true that the process may involve seeking explanations for various findings, but the correct diagnosis is not correct because it explains anything, but because criteria for making the diagnosis have been met. The correct diagnosis will provide, I argue, the "best" explanation for the diagnostician. Consider the following case (Lalazar et al. 2014):

A previously healthy 25-year-old man was admitted to the hospital because of abdominal pain, nausea, vomiting, and weight loss. Two weeks earlier, fever, chills, and weakness had developed. On examination, he was afebrile, had a slightly enlarged spleen and liver, and tenderness in the upper abdomen. Initially, the diagnostic possibilities that were considered included viral or bacterial infection, pancreatitis, and lymphoma.

Blood test results favored a viral infection, and antibodies in the blood to cytomegalovirus (CMV) were detected. Hypoalbuminemia (low albumin) was also present. A computerized tomographic scan of the abdomen showed enlarged gastric folds. At this point, diagnostic hypotheses were focused on those that were associated with enlarged gastric folds, which are uncommon. The serologic findings for acute CMV infection, along with the large gastric folds and hypoalbuminemia, were suggestive of Ménétrier's disease; other possibilities that were considered included gastric carcinoma.

A gastroscopy (direct visualization of the inside of the stomach through a fiberoptic endoscope) showed the enlarged folds; also noted was erosive gastritis, a type of inflammation. No evidence of cancer was seen. A biopsy showed viral inclusion bodies in several gastric mucosal cells that stained positive for CMV. The authors concluded that,

> ... the finding of inclusion bodies in the gastric mucosal cells that stained positively for CMV confirmed the diagnosis of CMV-associated Ménétrier's disease ... This diagnosis reasonably explains all the patient's presenting features: a prodrome of viral infection, splenomegaly, abdominal pain, erosive gastritis, large gastric folds, and hypoalbuminemia (Lalazar et al. 2014, 1348).

In this case, it seems clear that the diagnosis chosen as the correct one was done so *not* primarily because of explanatory considerations. But, as the authors note, the diagnosis "reasonably explains" all the patient's presenting findings. And, as I have argued, the correct diagnosis supplies the basis for the best explanation in this context.

In conclusion, there can be no doubt that the search for explanations motivates much scientific inquiry. And Lipton is surely correct when he asserts that the search for explanations illuminates the process of scientific discovery. Why is the world the way it is? In clinical medical science, much of the desire for explanation is for causal explanation, and many of the hypotheses formulated and tested are causal hypotheses. A causal hypothesis that is well confirmed by evidence must surely be different from any causal explanations stemming from it. Clearly, if the best explanation (e.g., the patient has measles) for the evidence (e.g., a rash) is the best-confirmed diagnostic hypothesis (e.g., the patient has measles), then we have nothing more than a tautology. Achinstein, who has construed explanation rather narrowly in his theory of evidence as being an objective, noncontextual relation between hypothesis and evidence, states that, "Explanation is a richer, more demanding idea than mere entailment" (2001, 149). Everywhere hypotheses and explanations are considered to be different. It seems much more plausible to view explanations as being based on hypotheses, since we know that a single hypothesis can potentially supply many different explanations, depending on the context in which an explanation is sought. Thus it appears reasonable to conclude that, at least in clinical medical science, confirmation is prior to explanation, and that explanation is not necessary for confirmation.

References

Achinstein, Peter. 1983. *The nature of explanation*. Oxford: Oxford University Press.
———. 2001. *The book of evidence*. Oxford: Oxford University Press.
———. 2005. Four mistaken theses about evidence, and how to correct them. In *Scientific evidence. Philosophical theories and applications*, ed. Peter Achinstein, 35–50. Baltimore: Johns Hopkins University Press.
Brown, Byron Wm, Jr., and Myles Hollander. 1977. *Statistics: A biomedical introduction*. New York: Wiley.
Earman, John. 1992. *Bayes or bust? A critical examination of Bayesian confirmation theory*. Cambridge: MIT Press.
Glymour, Clark. 1980. *Theory and evidence*. Princeton: Princeton University Press.
Harman, Gilbert. 1973. *Thought*. Princeton: Princeton University Press.
Harman, Gilbert. 1965. The inference to the best explanation. *Philosophical Review* 74: 88–95.
Hempel, Carl G. 1966. *Philosophy of natural science*. Englewood Cliffs: Prentice-Hall.
Howson, Colin, and Peter Urbach. 2006. *Scientific reasoning. The Bayesian approach*. 3rd ed. Chicago: Open Court.
Ladyman, James, Igor Douven, Leon Horsten, and Bas van Fraassen. 1997. A defence of van Fraassen's critique of abductive inference: Reply to Psillos. *The Philosophical Quarterly* 47: 305–321.
Lalazar, Gadi, Victoria Doviner, and Eldad Ben-Chetrit. 2014. Unfolding the diagnosis. *New England Journal of Medicine* 370: 1344–1348.
Lehrer, Keith. 1974. *Knowledge*. Oxford: Clarendon Press.
Lipton, Peter. 1991. *Inference to the best explanation*. New York: Routledge.
Psillos, Stathis. 1996. On Van Fraassen's critique of abductive reasoning. *The Philosophical Quarterly* 46: 31–47.
Sober, Elliott. 2008. *Evidence and evolution: The logic behind the science*. Cambridge: Cambridge University Press.
Suppes, Patrick. 1966. A Bayesian approach to the paradoxes of confirmation. In *Aspects of inductive logic*, ed. Jaakko Hintikka and Patrick Suppes, 198–207. Amsterdam: North-Holland.
Talbott, William. 2015. *Bayesian epistemology. Stanford encyclopedia of philosophy*, Summer 2015 ed, Edward N. Zalta. http://plato.stanford.edu. Accessed 5 Dec 2015.
Thagard, Paul R. 1978. The best explanation: Criteria for theory choice. *The Journal of Philosophy* 75: 76–92.
Van Fraassen, Bas C. 1980. *The scientific image*. Oxford: Clarendon Press.
———. 1989. *Laws and symmetry*. Oxford: Clarendon Press.
Yerushalmy, J., J.T. Harkness, J.H. Cope, and B.R. Kennedy. 1950. The role of dual reading in mass radiography. *The American Review of Tuberculosis* 61: 443–464.

Chapter 4
Confirmation of Hypotheses in Clinical Medical Science

Abstract In this chapter I discuss the ways in which information is gathered and used to confirm typical hypotheses encountered in clinical medical science. I discuss three kinds of hypotheses, namely, *therapeutic*, *etiologic*, and *diagnostic*. Therapeutic hypotheses are those concerned with treatments or other interventions, and etiologic hypotheses are those concerned with disease causation. Diagnostic hypotheses are those considered by clinicians when making a diagnosis. Examples from the medical scientific literature are extensively used. Included are an example of a randomized clinical trial and an N of 1 study for therapeutic hypotheses, and cohort, case – control, and cross – sectional studies are used for etiologic hypotheses. Approaches to the confirmation of diagnostic hypotheses are illustrated using actual published cases with discussions of the various strategies that are employed. Some possible pitfalls that may occur in the confirmation process are briefly discussed.

4.1 General

How are hypotheses confirmed in clinical medical science? How is evidence used in the confirmation process? In this chapter several different hypotheses are considered with the aim of exploring the various methods and strategies employed. They are illustrated using actual cases drawn from the medical literature.

For present purposes, hypotheses in clinical medical science can be divided into three groups: *therapeutic, etiologic, and diagnostic*. Therapeutic hypotheses are concerned with treatments or other health interventions. Etiologic hypotheses are concerned with the causes of diseases or other adverse health outcomes. Diagnostic hypotheses are those entertained by clinicians when making a diagnosis.

It may be observed that among the study designs that I discuss in this chapter, I place considerably more emphasis on those that address etiology, which are mostly observational (cohort, case-control, and cross-sectional), than therapy, which are mostly experimental. It seems to me that observational studies and their design and interpretation have received much less attention than deserved in the literature on evidence, and this emphasis is intended to help remedy this shortcoming.

© Springer Nature Switzerland AG 2020
J. A. Pinkston, *Evidence and Hypothesis in Clinical Medical Science*, Synthese Library 426, https://doi.org/10.1007/978-3-030-44270-5_4

4.2   Therapeutic Hypotheses

In clinical medical science, typical therapeutic hypotheses might include hypotheses such as:

H_0: Treatment A and treatment B give equivalent results
H_1: Treatment A is better than treatment B

Here, "treatment" could include a wide range of interventions, for example, drug therapy, radiation therapy, physical therapy, or some surgical procedure. The study could be primarily *observational*, or primarily *experimental*. In observational studies, researchers typically do not actively influence the treatment; rather, they observe treatments and make comparisons. An example would be the retrospective review of the results of two surgical procedures for the same condition at a particular institution over some time period.

In experimental studies, researchers may design an experiment using human subjects to test the effectiveness of some treatment, for example, to test whether some treatment is better than another treatment for the same condition, or better than no treatment at all. Controlled clinical trials and the N of 1 trial are examples of experimental studies.

4.2.1   Controlled Trials

An example of a controlled trial is that conducted by the Gastrointestinal Tumor Study Group to assess the value of adjuvant therapy following surgery for rectal cancer (Thomas and Lindblad 1988). The mainstay of treatment for rectal cancer is surgical extirpation of the tumor, usually consisting of either resection of the affected portion of the rectum with re-attachment end–to-end, or abdominal perineal resection, in which the rectum and anus are removed, with resultant colostomy. The results of several studies assessing the use of radiation therapy, chemotherapy, or both, had suggested that the use of these modalities as adjuvant therapy to surgery might improve outcomes.

The study was designed as a four-arm RCT of patients with stage B_2 (tumor penetrating deeply into the rectal wall) or stage C (tumor spread to nearby lymph nodes) adenocarcinoma of the rectum following "curative" resection (i.e., no clinical, radiological, surgical, or pathological evidence of disease remaining). Patients were randomly assigned to one of four groups: no adjuvant therapy (control), chemotherapy only, radiotherapy only, or radiotherapy and chemotherapy (combined modality).

Eleven cancer centers participated, 10 in the U.S. and one in Italy. Accrual to the study commenced in mid 1975, and by early 1980, 227 patients had been entered. At that point, the study was terminated since one of the arms showed a statistically higher recurrence rate than one of the other arms over a period of 18 months and

Table 4.1 First recurrence among treatment arms

	Control	Chemotherapy	Radiotherapy	Combined modality
No. of patients	58	48	50	46
No. of recurrences	32 (55%)	22 (46%)	24 (48%)	15 (33%)
Locoregional only	12	9	9	3
Locoregional and distant	2	4	1	2
Distant only	18	9	14	10

Source: Thomas and Lindblad (1988). p. 249, table IV

three separate interim analyses. On review, seven patients were declared ineligible and another 18 patients were withdrawn from the study after randomization but before treatment, leaving 202 patients available for analysis. Outcome variables studied were recurrences, disease-free survival (alive without evidence of recurrence), and survival (alive with or without recurrence). Statistical analysis was performed using the Kaplan-Meier product limit method (Kaplan and Meier 1958) and the log-rank test and the Cox proportional hazard method (Cox 1972).

The analysis presented is from mid 1987, nearly 6.5 years after the last patient had been entered into the study. At that time, 96 patients had demonstrated a recurrence (46%). The distribution of recurrences by type of recurrence is shown in Table 4.1:

Locoregional recurrences are those located in the area of surgery and radiotherapy, and distant sites are more removed, e.g., lungs and liver. A proportional hazard analysis adjusted for stage and type of surgery showed differences among the four treatment arms to be significant ($p = .04$). An adjusted comparison of time to recurrence between the control group and the combined modality therapy group showed a significantly longer disease-free interval for the combined modality group ($p = .005$). The difference in survival between the combined modality therapy group and the control group was significant ($p = .01$). Importantly, the authors stated, "In addition, the importance of a no adjuvant therapy control arm cannot be overemphasized. Our results for this cohort of patients are much better than would be suggested by historical controls" (Thomas and Lindblad 1988, 250).

Although the number of patients was relatively small, the Gastrointestinal Tumor Study Group trial was one of the first RCTs that showed a significant difference in favor of combined modality adjuvant therapy in rectal cancer, and similar trials continue to be conducted with the aim of improving treatment outcomes with the use of newer cancer chemotherapy drugs and improved surgical and radiotherapy techniques.

4.2.2 N of 1 Trials

The routine treatment of a patient resembles in many respects an experiment, in the sense that a treatment is prescribed based on a diagnostic hypothesis, and if, as hoped and predicted, the patient improves, it is concluded that the treatment was efficacious. But, as Guyatt et al. (1986, 889) point out, this conclusion can be wrong for a number of reasons:

1. The patient's illness may simply have run its course, and recovery would have occurred with no treatment;
2. The patient's symptoms, signs, or laboratory values at presentation may represent temporary extreme levels that will "regress toward the mean" when they are next measured; thus any treatment begun between the two measurements will appear to be effective;
3. The "placebo effect," which has been said to be responsible for as much as 30% of many treatment effects, may underlie the improvement;
4. When both clinician and patient know what is expected from the treatment, their lack of "blindness" may influence their interpretations of whether the symptoms and signs of illness have been relieved, and
5. When the patient appreciates the efforts of the clinician, a willingness to please (or at least not to disappoint) the clinician may cause the patient to minimize symptoms or overestimate recovery (the "obsequiousness bias").

Although treatments are often based on sound scientific studies like clinical trials, many times, arguably the majority, treatment decisions cannot be made on their basis. First, no controlled trial may be available; or if studies *are* available, their results or the interpretation of their results may be conflicting. Even when a randomized trial has generated a positive result, it may not apply to an individual patient. For example, the patient may not meet the eligibility criteria and thus generalization of findings may not be appropriate.

The N of 1 study ("N" being a standard abbreviation for sample size) has been suggested as a possible method for testing the hypothesis of drug efficacy in an individual patient. The method has a history of several decades of use in experimental psychology to investigate behavioral and pharmacologic interventions. Ideally, an easily determined and reliable measure of treatment efficacy (a "target") should be available, and should be a symptom or sign that is troubling to the patient. As Guyatt et al. (1986, 890) state, ". . . rapid improvement must occur when effective treatment is begun, and the improvement must regress quickly (but not permanently) when effective treatment is stopped. Selecting signs and symptoms that are particularly troubling or relevant to the individual patient provides one of the major advantages of the N of 1 randomized controlled trial over conventional controlled trials, in which tailoring of outcomes is generally sacrificed in favor of uniform end points that are applied to all study participants."

Although several study designs have been found useful, Guyatt et al. employed a pair design in which an active drug was compared with a placebo. Each pair consisted of two treatment periods, one period in which active drug was used, and the other in which placebo was used. For each pair, whether active drug or placebo came first was decided randomly, and both patient and clinician were blinded. Identical appearing active drug and placebo were prepared by the pharmacy, which also was responsible for the random allocation. The patient reported relevant symptoms or signs according to a standard form for each treatment period.

Table 4.2 An N of 1 randomized controlled trial of theophylline

Symptom	Pair 1		Pair 2	
	Period 1 (Drug)	Period 2 (Placebo)	Period 1 (Drug)	Period 2 (Placebo)
	Score[a]			
Shortness of breath	3	6	3	6
	3	5	3	5
	4	7	4	5
Need for inhaler	3	5.5	3	5
Sleep disturbance	5	5.5	3	5

Source: Guyatt et al. (1986). p. 890; table 1. Copyright © 1986 Massachusetts Medical Society. (Reprinted with permission from Massachusetts Medical Society)
[a]The patient rated his symptoms on a 7-point scale in which 7 represented optimal function and 1 represented severe symptoms

The results of a double blind N of 1 randomized trial of theophylline use as part of a multidrug regimen in a 65-year-old man with severe asthma are shown in Table 4.2:

Each treatment period lasted 10 days, and at the end of each period the patient rated his symptoms on the 7-point scale described in Table 4.2. As Guyatt et al. (1986, 890) describe, "The patient reported that he was feeling much worse during the first period of each of the two pairs completed, and both he and his physician felt confident that the Period-2 treatments were superior. At that point it was decided that it would be unfair to ask the patient to undergo another pair of blinded treatment periods. When the code was broken, it turned out that the patient was receiving active drug during the first periods of both pairs and placebo during the second. The drug was discontinued and the patient felt considerably better. In retrospect, the theophylline was probably contributing to nocturnal gastroesophageal reflux and pulmonary aspiration."

The authors note that often results may be evident in the absence of statistical analysis. If, however, analysis is performed, in the simple case where each period results in a simple preference for one of two alternatives (e.g., drug or placebo), one can use the binomial distribution to estimate the probability of obtaining the observed results. Often quantitative results will be obtained that allow a more powerful analysis such as a paired t-test or the Wilcoxon test to be performed.

In deciding whether to execute an N of 1 trial, Straus et al. (2011, 132) advise that "The crucial first step... is to have a discussion with the patient to determine his interest, willingness to participate, expectations of the treatment, and desired outcomes." They have advanced guidelines for these and other considerations.

4.3 Etiologic Hypotheses

Another type of hypothesis in clinical medical science is that involving disease causation. For example,

H_2: Exposure to x causes outcome y

Here, persons exposed to some x, say tobacco smoke in cigarette smokers, are hypothesized to be more likely to get outcome y, say lung cancer, than persons not exposed (e.g., nonsmokers or ex-smokers). Most such studies are observational, since they involve exposure to some noxious agent that is thought to cause disease. To confirm such hypotheses, persons both exposed and non-exposed are compared with regard to the frequency of outcome y. Cohort studies, case-control studies, and cross-sectional studies are examples of types of observational studies that address disease causation.

4.3.1   Cohort Studies

Cohort studies are observational studies of one or more groups of (usually) non-diseased persons exposed to some agent believed to cause disease or some other outcome. Persons that have been exposed to the agent, often at more than one level or intensity of exposure, are assembled and followed over some time period, often years or even decades, and the frequency of disease or other outcome is recorded. Often one large group comprises the cohort, which includes individuals unexposed as well as others exposed at various levels of exposure. If exposure is unrelated to the subsequent development of disease, then the frequency of disease in persons at each exposure level, including the unexposed, would be expected to be similar.

For example, one well-known large cohort study was conducted in the United Kingdom in which mortality (the outcome measure) was assessed in relation to cigarette smoking. In 1951, smoking information was obtained from about two-thirds (34,439) of all male British doctors, and updated over five decades through 2001 (Doll et al. 2004). Interim analyses were reported at 4, 10, 20, and 40 years of follow-up. Overall mortality (death from any cause) and cause-specific mortality (death from a particular cause) were assessed in different cohorts that were assembled according to smoking status. The statistical hypotheses utilized ratios of the mortality experience of smokers with non-smokers overall, and the various levels of smoking (heavy, moderate, and light), were also compared with non-smokers. Standardized survival curves were constructed and mortality ratios were standardized to age and other factors (known as standardized mortality ratios) and assessed using chi-square tests and tests for trend. Overall mortality, as well as cause-specific mortality, was strongly related to cigarette smoking ($p < .0001$).

Among the specific causes of death that were analyzed, lung cancer, chronic obstructive lung disease, and ischemic heart disease were the most strongly associated with smoking, but all of the other conditions analyzed were also strongly associated, including cancers of the mouth, pharynx, larynx, esophagus, all other neoplasms as a group, and vascular disease. Overall mortality, as well as cause-specific mortality, showed a dose-response relationship: among smokers, heavy smokers had the highest risk, whereas light smokers had the lowest risk. Interim reports also showed the significant risk and were instrumental in demonstrating that the relationship between smoking and deleterious health effects was not mere coincidence, but causal (Greenhalgh 2010, 41).

Although it is a cohort study, several considerations lead to the conclusion that possible sources of bias or imprecision have been minimized and that the evidence favoring the hypothesis that cigarette smoking causes a number of adverse health conditions is convincing. The subjects in the study, male British doctors, are easily traceable because the British Medical Association knew their addresses. Their responses to questionnaires are thought to be generally accurate because of their educational level and because health information was sought in addition to smoking habits. The 1978 questionnaire, for example, also requested information on a wide range of characteristics including alcohol use, height, weight, blood pressure, and medical history. Recorded causes of death are believed to be relatively accurate since the doctors' medical conditions were probably well known to their treating physicians.

The authors concede that the study has some degree of bias and confounding, however. For example, confounding by alcohol consumption can elevate risks associated with cigarettes, including various cancers and cirrhosis of the liver, but alcohol consumption also may act in the opposite direction to decrease risk of ischemic heart disease and perhaps of some other conditions. Because of the large number of subjects under study (over 34,000 initially) and the strengths of the associations, however, the authors concluded that the study has provided strong evidence of a causative role by cigarette smoking on overall mortality as well as the specific causes of death under study among the cohort.

4.3.2   Case-Control Studies

In a case-control study, cases of a disease or other health outcome are assembled and various exposures or other possible risk factors are recorded. One or more control groups that consist of persons without the disease or other health outcome are selected, and similar information concerning exposures or the presence of other risk factors is likewise obtained. By comparing the exposure and risk factor histories of cases and controls, differences of possible etiologic interest are sought (Rothman et al. 2008).

Case-control studies commonly use the *exposure odds ratio* as an estimate of the relative risk of disease among the exposed compared with the non-exposed. It is the ratio of the odds of exposure among cases to the odds of exposure among controls for some exposure or other factor. A high odds ratio means that exposure is more frequent among cases than controls. The odds ratio in a case-control study is an estimate of the incidence rate ratio or risk ratio obtained in a cohort study. Statistical analyses are conducted most often using chi-squared tests such as the Mantel-Haenszel technique (Mantel and Haenszel 1959) or logistic regression (Hosmer and Lemeshow 1989).

Under ideal circumstances, a case-control study may be thought of as yielding the same information that would be obtained in a cohort study. However, in general, case-control studies are known to be more subject to biases than are cohort studies. One problem is the selection of controls. Cases may be selected in a variety of ways, but study validity is threatened if the control group is not drawn from the same population that gave rise to the cases. Imagine, for example, a case-control study of Hodgkin's disease (also called Hodgkin lymphoma) carried out at the Stanford University Medical Center. Patients with Hodgkin's disease come to Stanford from all over the world. Although numerous cases would be available for inclusion in the study, in order to obtain controls, how would one determine the specific source population that gave rise to the cases?

Case selection may also be problematic, in that ideally cases would consist of a direct sampling of cases within a source population. Not *all* cases need to be included, however. Cases, like controls, can be randomly selected from the source population, as long as the sampling is independent of any exposure under study. Thus, cases that are identified in a single institution or practitioner's office are candidates for possible inclusion in a case-control study. For any particular disease, the source population for the cases treated in the clinic is all the people that would attend that clinic or office and would be diagnosed with the disease if they had the disease in question.

In a case-control study, as noted, a source population is defined and ideally the selection of controls will involve direct sampling from that population. Two basic rules that have been advanced are (Rothman et al. 2008, 116):

1. Controls should be selected from the same population – the source population – that gives rise to the study cases. If this rule cannot be followed, there needs to be solid evidence that the population supplying controls has an exposure distribution identical to that of the population that is the source of cases, which is a very stringent demand that is rarely demonstrable.
2. Within strata of factors that will be used for stratification[1] in the analysis, controls should be selected independently of their exposure status, in that the sampling rate for controls ... should not vary with exposure.

As indicated, these ideal circumstances are rarely met and in practice it is most often necessary to select controls in some other way. One method is to use *neighborhood* controls. So, for example, a control may be selected who resides in the same residential neighborhood as the case for which the control was selected. This may easily introduce bias, however, in that neighbors of cases may not be representative of the opportunity for exposures characteristic of cases. To illustrate, suppose cases were obtained from a single U.S. Veterans Administration hospital. Controls selected from neighbors may differ considerably in their exposure histories. For

---

[1] The *stratification* referred to is a common method for attempting to control bias such as confounding. For example, by choosing controls with the same age structure (strata) as cases, any influence due to age imbalance would be minimized or eliminated.

example, only a minority may have served in the military with their attendant exposures secondary to combat or weapons handling, as would be the case with, say, veterans of the Vietnam conflict that had been exposed to Agent Orange. Several other methods of control selection have also been proposed, including random-digit (telephone) dialing, friend controls, or general population controls. All of these methods have potential biases that pose a threat to study validity.

When there is a strong association between an exposure and a particular disease, however, the aforementioned limitations of the case-control study design may sometimes be easily overcome. This was apparently true of the case-control studies of the relation between cigarette smoking and lung cancer that preceded the cohort study by Doll et al. (2004) of male British doctors described above. Several case-control studies from Western Europe (e.g., Doll and Hill 1950) and North America (e.g., Wynder and Graham 1950) led in 1950 to the conclusion that smoking was a cause of lung cancer (Doll et al. 2004, 1), which in turn led to the decision to launch the British doctor study.

In the case-control study by Doll and Hill (1950), for example, lung cancer cases were drawn prospectively from several London hospitals between April, 1948 and October, 1949. For each case, a control patient of the same sex and same five-year age group was selected concurrently, and for the most part was selected from the same hospital. All patients were personally interviewed. Cases of lung cancer totaled 709, of which 649 were men and 60 were women. Among men, 0.3% of cases and 4.2% of controls were non-smokers. For women, the corresponding figures were 31.7% and 53.3%. Statistical tests of significance on these differences were $p < .0001$ for men and $.01 < p < .02$ for women.

Among smokers, a relatively high proportion of lung cancer cases tended to be heavier smokers. Lung cancer patients were observed, on the whole, to have begun smoking earlier and continued for longer than controls, but these differences were not statistically significant. It will be noted that the percentages of smokers among male and female controls were 95.8 and 46.7, respectively. These high percentages no doubt reflect, at least partially, the over-representation of smoking related ailments associated with hospitalized patients, since today we know that cigarette smoking is associated with many diseases including various lung ailments and heart disease. Thus, even with the heavily biased control group, a relationship between cigarette smoking and lung cancer was nevertheless found.

In the case-control study reported by Wynder and Graham (1950), 605 male cases of histologically verified epidermoid, undifferentiated, or unclassified carcinomas of the lung were identified from a geographically diverse group of hospitals and medical practices in the U.S. Control subjects numbered 780 males that were drawn from hospitalized patients on the general medical and surgical services of three of the hospitals from which cases were drawn. All subjects completed the same questionnaire and over 90% were personally interviewed. Data sought included lung disease in general, histories of tobacco and alcohol use, and occupational history. The amount of cigarette use was divided into six categories, including one for non-smokers. The five categories of cigarette use ranged from minimal smoking (1–9 cigarettes per day for at least 20 years) to chain smokers (≥ 35 cigarettes per day for at least 20 years). Adjustments were made for persons that had smoked for less than 20 years, although the great majority had smoked for 20 years or longer.[2] The age distribution of the controls was adjusted to that of the cases.

Statistical analysis of the data revealed a strong association between cigarette smoking and lung cancer ($p < .0001$). Among their observations was that the occurrence of carcinoma of the lung in a male nonsmoker or minimal smoker was a rare phenomenon (2%).

A case-control study that investigated a possible relationship between cigarette smoking and myocardial infarction in healthy young women was conducted in the northeastern United States during the mid 1970s (Slone et al. 1978). Trained nurse interviewers conducted interviews with all cases and controls, which were drawn from 152 participating hospitals that had coronary care units. All cases met World Health Organization criteria for "definite myocardial infarction." Potential controls as close in age as possible to cases were selected from the surgical, orthopedic, and medical services of the same hospital.

Since the object of the study was to examine factors related to myocardial infarction in otherwise healthy young women, known risk factors for myocardial infarction and ischemic heart disease were excluded. Disqualifying factors included drug-treated hypertension, drug-treated diabetes mellitus, previous myocardial infarction, drug-treated obesity, drug-treated angina pectoris, abnormal blood lipids, and use of oral contraceptives within the month prior to admission.

To ensure that the comparisons would not be confounded, each case of myocardial infarction was matched to four controls that were for the most part from the same five-year age group, and from the same hospital (41%), or, failing that, from the same area of residence (59%). Results from an analysis of 55 cases and 220 controls are shown in Table 4.3. Of the cases, 89% were smokers compared with 55% of the controls.

In this analysis, the relative risk of myocardial infarction among current smokers compared to those who are not current smokers is estimated by the exposure odds ratio, which is the ratio of the odds of exposure (current smoking) among cases to the odds of exposure among controls:

Exposure odds for cases: 49/6   Exposure odds for controls: 120/100
Odds Ratio = (49/6)/(120/100) = 6.8 (chi square 22.2; $p < .001$)

With women who have never smoked as the reference category (relative risk set at 1.0), the estimate of the relative risk for ex-smokers (women that had not smoked for at least 1 year) is 1.4, as seen in Table 4.4:

As can be seen, the relative risk for myocardial infarction increases with an increase in the number of cigarettes smoked per day. The relative risks associated

---

[2]For example, a person who had smoked 20 cigarettes daily for 10 years was classed as having smoked 10 cigarettes daily for 20 years.

Table 4.3 Relation of myocardial infarction to smoking habits in 55 cases and 220 controls

	Cases	Controls	
Current smokers	49	120	169
Not current smokers	6	100	106
	55	220	275

Source: Adapted from Slone et al. (1978). p. 1275, table 3. Copyright © 1978 Massachusetts Medical Society. (Reprinted with permission from Massachusetts Medical Society)

Table 4.4 Relation of myocardial infarction to smoking habits in 55 cases and 220 controls

	Cases		Controls		
Cigarettes/day	no.	%	no.	%	Estimated relative risk
Never smoked	4	7	73	33	1.0[a]
Ex-smoker	2	4	27	12	1.4
1–14	8	15	33	15	4.4
15–24	15	27	59	27	4.6
25–34	12	22	16	7	14
> 35	14	25	12	5	21

Source: Slone et al. (1978). p. 1275, table 3. Copyright © 1978 Massachusetts Medical Society. (Reprinted with permission from Massachusetts Medical Society)
[a]Reference category

with the categories of one to 14, 15–24, 25–34, and 35 or more cigarettes per day are 4.4, 4.6, 14, and 21, respectively. A statistical test for trend showed that the trend of increasing relative risk with increasing cigarette consumption is statistically significant ($p < .001$).

Although the authors could not evaluate duration of cigarette smoking in their study, they observed that cigarette smokers who died of conditions other than coronary heart disease have been found at autopsy to have thickening of the inside lining of the coronary arteries as well as an excess of atheromatous plaques, suggesting that duration of smoking may have a cumulative deleterious effect. They also noted that cigarette smoking might exert a precipitating effect, since the relative risk in ex-smokers (1.4) represented only a small increase over non-smokers.

In perhaps most case-control studies, however, an exposure-disease relationship is weaker or non-existent and the potential for biases to exert themselves may pose a significant difficulty in interpretation of results. An example is provided by the case-control studies that investigated a possible increase in risk of Hodgkin's disease among persons that had undergone tonsillectomy in childhood (Mueller 1996). Tonsillectomy is a risk factor for multiple sclerosis and poliomyelitis, both of which share epidemiological characteristics with Hodgkin's disease. It was postulated that Hodgkin's disease might have an infectious origin, in view of its association with fever, night sweats, and cervical adenopathy. Since the disease usually starts in the neck with enlarged lymph nodes, the removal of lymphoid tissue in the tonsillar and adenoid areas such as occurs with tonsillectomy might remove a lymphoid barrier to an as yet unidentified infectious agent, perhaps a virus, allowing access to cervical lymph nodes and subsequent spread. In 14 published studies, however, the relative risk of persons with prior tonsillectomy relative to those without has ranged from 0.7 to 3.6.

The reasons for this degree of variability are unclear. Hodgkin's disease has previously been shown to be related to socioeconomic status, with children in relatively poor living conditions more susceptible to the disease, while among young adults and middle aged persons evidence exists that Hodgkin's disease may be a host sequela to a common infection. Tonsillectomy rates in the populations studied varied widely, from 9% in Denmark to 74% among Boston area cases (Mueller 1996, 900). Tonsillectomy rates are also directly related to socioeconomic status as well as local medical practice (Gutensohn et al. 1975, 22).

In their study of tonsillectomy and Hodgkin's disease, Gutensohn et al. (1975) used two comparison groups, siblings and spouses, to control for socioeconomic status in childhood and adulthood, respectively. The study compared 136 young adult patients with Hodgkin's disease being followed at the Joint Center for Radiation Therapy, Harvard Medical School, in 1972 with their 315 siblings and 78 spouses. All subjects completed a questionnaire. Matched analyses were used in all comparisons since controls were matched to cases for factors that might correlate with tonsillectomy. Risk ratios expressing the risk of Hodgkin's disease were estimated for persons who have had a tonsillectomy in relation to a risk of unity for those who have not. On the basis of the case-spouse comparison, the risk ratio of Hodgkin's disease among persons with previous tonsillectomy was 3.1 ($p < .05$) and on the basis of the case-sibling comparison it was 1.4, which was statistically non-significant. The investigators concluded that any relation between Hodgkin's disease and tonsillectomy is either non-causal or is complex and modified by family size.

A later, larger population-based case-control study of tonsillectomy and Hodgkin's disease from the Detroit and eastern Massachusetts area that studied 556 cases and 1499 siblings found no significant elevated risk of Hodgkin's disease among young adults (age 15–39 years) or middle aged adults (age 40–54 years), with risk ratios of 1.0 and 1.5, respectively. Among older persons, the risk ratio was significantly elevated, 3.0 (95% confidence interval 1.3–6.9), but the data were sparse. The authors concluded that it was unlikely that prior tonsillectomy is a risk factor for the development of Hodgkin's disease in young or middle aged adults, but whether it is a risk factor for the malignancy occurring late in life is unclear (Mueller et al. 1987).

For rare diseases, however, the case-control study design may be the only feasible method for exploring exposure-disease relationships. Although sporadic case reports of rare conditions often provoke curiosity, an unusual clustering of cases of some rare condition in a relatively small geographic area may cause concern and provide an opportunity for a meaningful investigation. This occurred, for example, in the Boston area with the observation of several cases of a rare vaginal cancer in young women (Herbst et al. 1971).

Cancer of the vagina is rare. Most cases are of squamous cell histology, and occur in women over 45 years of age, with a median age at diagnosis of 69 years (Daling

Table 4.5 Summary of data comparing patients with matched controls

	Cases	Controls	Chi-square	P value
Bleeding in this pregnancy	3/8	1/32	4.52	<.05
Any prior pregnancy loss	6/8	5/32	7.16	<.01
Estrogen given in this pregnancy	7/8	0/32	23.22	<.00001

Source: Adapted from Herbst et al. (1971). p. 879, table 2. Copyright © 1971 Massachusetts Medical Society. (Reprinted with permission from Massachusetts Medical Society)

and Sherman 1996, 1117–1118). Between 1966 and 1969, however, seven young females 15–22 years of age with clear cell or endometrial type adenocarcinoma of the vagina were seen at Vincent Memorial Hospital in Boston. These seven cases, plus another case of clear cell adenocarcinoma in a 20-year-old female patient treated in 1969 at another Boston area hospital, formed the case series for a case-control study to identify possible risk factors.

Four matched controls were selected for each case, based on the birth records of the hospital in which each case was born. Controls were born within 5 days from when the case was born, and on the same type of service (ward or private). All mothers and daughters were personally interviewed using a standard questionnaire by trained interviewers. Statistical tests included chi-square tests, paired and unpaired t-tests, and a matched control method for nonparametric data.

Information on numerous factors was analyzed, but only three were statistically significant, as shown in Table 4.5:

Bleeding during pregnancy and a history of prior pregnancy loss are associated with high-risk pregnancies, which provided the indication for stilbestrol (a type of estrogen) administration. All of the mothers using stilbestrol began using it during the first trimester of pregnancy.

To estimate the frequency of stilbestrol administration and the risk of development of these tumors in female offspring whose mothers took stilbestrol during pregnancy, the authors studied deliveries from a special high-risk pregnancy clinic at Boston Lying-In Hospital that occurred from 1946 to 1951, the period during which the eight cases were delivered. Of approximately 14,500 ward deliveries, stilbestrol was administered to 675, about one in 21. Only one case of vaginal adenocarcinoma in a young female is known to the authors to have occurred from deliveries at that hospital, so the risk to exposed offspring is thought to be low. The authors also note that sporadic cases of adenocarcinoma of the vagina occur in young females whose mothers were not exposed to stilbestrol during pregnancy, which was also observed in one of the eight cases in the present study. Thus factors other than exogenous maternal estrogen use during pregnancy appear to be involved.

4.3.3 Cross-Sectional Studies

Studies that include all persons in a population, or some representative sample of all such persons, at a point in time or over some relatively short time interval are usually referred to as *cross-sectional studies*. In general, subjects are not selected based on exposure or disease status, but information on these factors, in addition to other factors deemed of importance to the investigators, are gathered.

Cross-sectional studies may be employed for a variety of purposes, including estimating the prevalence of disease in a population or gathering other information for use by health planning agencies. However, they are also frequently performed to assess exposure – disease relationships with aims similar to cohort and case-control studies. The cross-sectional study design is most often used to study risk factors for diseases of slow onset and long duration, and for which medical care is usually not sought until late in the course of the disease (Kelsey et al. 1986, 187). Hypertension, various mental disorders, osteoarthritis, and chronic bronchitis are examples of such conditions. Case-control studies of these diseases are often less practical and more difficult to interpret, since it is usually difficult to establish when someone becomes a "case." Most often, incident cases in case-control studies are identified when they seek medical care. Cohort studies suffer the same problem, in that it is difficult to say at what point in time people have these diseases and when they do not. Cross-sectional studies have one major advantage over many case-control studies since they are often based on a sample of the general population, and not just on individuals seeking medical care, and therefore their results may be more easily generalized. They also share the advantage over cohort studies possessed by case-control studies in that they are usually performed more quickly and at less cost.

Since information on exposure and disease status, as well as other factors, is usually not available prior to conduct of the study, techniques for control of confounding variables prior to analysis such as matching or stratification are likewise not available, and multivariable statistical techniques are often used for hypothesis testing. These methods are applicable to studies where the outcome is either the presence or absence of disease, or for continuously distributed variables such as blood pressure. For example, for a continuously distributed outcome variable, the multivariable model is an extension of simple linear regression, represented by the equation

$$y = a + bx \tag{4.1}$$

Here, "y" is the dependent, or outcome variable; "x" is an independent or predictor variable; and, "a" and "b" are constants estimated from the data.

The multivariable model is

$$y = a + b_1 x_1 + \ldots + b_r x_r \tag{4.2}$$

Here, "r" denotes the number of predictor variables included in the model, and where $x_1, \ldots, x_r$ are a particular person's set of values for the regressor variables, which can be binary or continuous.[3]

---

[3]Both equations contain a random error term that has been omitted.

An example of a cross-sectional study where the outcome of interest was blood pressure is the study of Japanese men by Ueshima et al. (1984), and discussed by Kelsey et al. (1986, 194–198). Between 1975 and 1977, these investigators surveyed 487 men aged 40–69 years from Osaka, an urban area, and 365 men of the same age from Akita, a rural area. Information on several variables thought to be related to blood pressure was obtained, including age, smoking history, alcohol intake, weight, height, hemoglobin, uric acid, cholesterol, and triglycerides. All of the factors were considered continuous variables. Alcohol consumption was the primary variable of interest and was categorized on a six-point scale ranging from total abstention to consumption greater than 83 grams per day. Smoking was measured on a five-point scale ranging from non-smokers to greater than 19 cigarettes per day.

Selected results of a stepwise multiple regression of systolic blood pressure on selected variables are shown Table 4.6:

The dependent variable for the above analyses is systolic blood pressure. The strongest predictor for systolic blood pressure for both areas was age. The coefficient for age in Osaka was 0.7187. The interpretation of this result is that, on average, in this population a one-year increase in age for one man is associated with a blood pressure that is 0.7187 mmHg higher than in another man, contingent on the remaining variables in the model being equal for the two men. Thus the estimate of a 20-year difference in age is associated with a difference in systolic blood pressure of $20 \times 0.7187 = 14.374$ mmHg.

The quantity denoted by ΔR^2 is the proportion of the total variation in systolic blood pressure that is accounted for by each of the variables in the regression equation. For example, a value close to unity means that most of the variability is

Table 4.6 Stepwise multiple regression[a] of systolic blood pressure on daily alcohol intake, ponderosity index, cholesterol, triglycerides, hemoglobin, uric acid, smoking and age, men aged 40–69 years, Osaka and Akita, 1975–1977

Variable	Coefficient	ΔR^2	T value	p value
Osaka (487 men)				
Age (year)	0.7187	0.0918	7.4801	<0.001
Daily alcohol intake	2.4704	0.0465	4.5260	<0.001
Ponderosity index[b]	1.8275	0.0328	3.8089	<0.001
Uric acid (mg/dl)	1.4627	0.0059	1.8526	NS
Akita (365men)				
Age	0.8208	0.0596	5.7017	<0.001
Daily alcohol intake	3.1268	0.0528	4.6465	<0.001
Ponderosity index[b]	1.2090	0.0135	1.6023	NS
Triglycerides (mg/dl)	0.0327	0.0080	1.7754	NS
Hemoglobin (g/dl)	1.5488	0.0053	1.6675	NS

Source: Adapted from Ueshima et al. (1984). p. 589, table 5
R^2 for Osaka and Akita are 0.1771 and 0.1458, respectively
[a]The selection criterion is F = 2
[b]Ponderosity index = weight/height$^3 \times 10^6$
NS nonsignificant

accounted for by that variable. For Osaka men, for example, age alone accounts for 9.18% (100 × .0918) of the variation in systolic blood pressure, and alcohol intake accounts for an additional 4.65%. Adding in the other two variables (uric acid and ponderosity index) gives a total of 17.71% of the variation in systolic blood pressure accounted for by the four variables. This low percentage indicates that most of the variability in systolic blood pressure is not accounted for by the variables selected for inclusion in the statistical model, and that most probably important determinants of systolic blood pressure were not measured.

Two major limitations of the cross-sectional study design have been recognized, namely, problems associated with the temporal relation of exposure to disease, and problems associated with measuring prevalence rather than incidence (Checkoway et al. 1989, 223–224). The temporal problem is that it is often not possible to be confident that exposure preceded disease, thus hindering causal inference. Consider the example of a study of the prevalence of cardiovascular disease among London bus drivers and conductors (Morris et al. 1953). Bus drivers, whose job is mostly sedentary, were found to have a higher prevalence of cardiovascular disease than conductors, whose job requires more physical activity. However, the drivers' cardiovascular risk factors, such as obesity, and perhaps symptom manifestations like shortness of breath on exertion, were probably factors that motivated them seeking jobs as drivers, rather than conductors, who have to physically exert themselves to a greater extent. This type of bias is always of concern in studies that elicit information on exposures or other risk factors simultaneously with disease.

A problem that arises when prevalence rather than incidence is measured is that prevalence is a function of incidence and duration of disease. Diseases that have a longer duration will tend to be over-represented in these studies, and often it is not possible to determine whether observed effects on disease prevalence are due to effects on incidence, duration, or both. Checkoway et al. (1989, 224), provide an example: Suppose that exposures in section A of an industrial plant cause fatal coronary heart disease, and exposures in section B cause non-fatal coronary disease. A cross-sectional study might reveal a higher prevalence of coronary heart disease in section B workers, even if the combined incidence of both fatal and non-fatal coronary heart disease were the same in both sections of the plant. This type of bias could be avoided in a full cohort study.

4.3.4 Remarks on the Challenges of Etiologic Research

Etiologic research, as we have seen, is concerned with identifying factors that are associated with disease occurrence, with the aim of identifying those factors that are causative. This is, more often than not, a complex process. Consider etiologic research on cancer as an example: the list of subjects for study is quite long, and includes tobacco, alcohol, ionizing radiation, solar radiation, electromagnetic fields and radiofrequency radiation, occupation, air pollution, diet and nutrition, infectious agents, and chemicals, to name a few. Thousands upon thousands of studies have been done, and many, if not most, putative causal inferences continue to generate controversy. We discussed earlier a possible association between tonsillectomy and Hodgkin's disease. As another example, consider studies on the etiology of pancreatic cancer. Anderson et al. (2006) listed over 50 possible risk factors that have been studied in relation to the etiology of pancreatic cancer. Here I will consider just two: tobacco use and coffee drinking.

The authors note that the most consistent risk factor for pancreatic cancer is cigarette smoking. In 2004, the International Agency for Research on Cancer (IARC) concluded: "Cancer of the pancreas is causally associated with cigarette smoking. The risk increases with duration of smoking and number of cigarettes smoked daily. The risk remains elevated after allowing for potentially confounding factors such as alcohol consumption. The relative risk decreased with increasing time since quitting smoking" (IARC 2004, 1182). What evidence is this conclusion based upon? Anderson et al. (2006) note that an IARC working group had concluded in 1986 that smoking was a cause of pancreatic cancer, and that conclusion was based on the evaluation of nine cohort studies and eight case-control studies, and that all of the cohort studies and all but one of the case-control studies showed increased risks for smokers and most had evidence of a positive dose response. The estimates of relative risk were about twofold, with some as high as sixfold in association with heavy cigarette use. They also noted that studies after the 1986 report supported the conclusion, including at least 10 cohort studies and 18 case-control studies. One hospital-based case-control study found no association, but the reported odds-ratios (male OR 1.42, 95% C.I. 0.73–2.78; female OR 1.12, 95% C.I. 0.51–2.45) were consistent with an increased risk from cigarette smoking (Clavel et al. 1989).

Attention was drawn to the possibility that coffee drinking was associated with an increased risk of pancreatic cancer with the publication in 1981 by MacMahon et al. of a case-control study showing a twofold to threefold increase in risk among coffee drinkers consuming three cups a day. The effect became evident after control for smoking. The high correlation between coffee drinking and smoking in the U.S. at that time made it difficult to be confident that the effect was not due to some residual confounding by smoking, since smoking itself is considered to be associated with pancreatic cancer.

In 1991, the IARC published its evaluation of the risk of coffee drinking and pancreatic cancer. The report evaluated 21 case-control studies and six cohort studies. It was concluded that the data as a whole were suggestive of a weak association between high levels of coffee drinking and pancreatic cancer, but that bias or confounding could account for the association. In their review of studies published since the 1991 IARC report, Anderson et al. note that 11 case-control studies found null or non-significant associations between coffee consumption and pancreatic cancer, and that five cohort studies also showed null associations. Nevertheless, they also noted that at least five studies have shown statistically significant associations. They conclude: "The possibility that coffee is associated with an increased risk of pancreatic cancer seems unlikely. Reports of an increased risk associated with coffee drinking likely result from residual confounding from cigarette smoking, and possibly from other sources of confounding or bias" (2006, 741).

Thus in etiologic research multiple studies are usually required even to establish relatively strong associations, and a considerably greater number is required when associations are weak or nonexistent.

4.4 Diagnostic Hypotheses

Diagnosis in clinical medical science has been said to mean "the identification of a disease by the investigation of its various manifestations" (Harvey 1994, v.). The term *differential diagnosis* refers to evaluating more than one diagnostic hypothesis that may be entertained in the process of making a diagnosis. This process may involve taking a medical history, performing a physical examination, collecting laboratory and imaging data, and performing other tests like electrocardiograms, among other things. At least some of this information may need to be obtained more than once during the diagnostic process, since a disease may be evolving in its manifestations.

From this body of information will come the evidence that will be used to confirm or disconfirm the various alternative diagnostic hypotheses. In the ideal case, what is sought is the correct diagnosis, or the true hypothesis, but due to the inherent uncertainty involved,[4] what usually emerges is a "working diagnosis," the hypothesis that is best confirmed. This working diagnosis will serve as the basis for further diagnostic testing or therapeutic intervention. A potentially complicating factor is that a patient may have more than one correct diagnosis, for example, a woman with diabetes, hypertension, and endometrial cancer existing contemporaneously.

How is evidence used to confirm or disconfirm diagnostic hypotheses? Becoming familiar with the process of advancing and confirming diagnoses is an important part of a medical education and one popular method for teaching these skills involves case presentations that employ diagnostic reasoning, including those such as clinicopathologic conferences that are published in medical journals. An illustrative case is that involving a patient with migrating polyarthralgias (migratory pains involving multiple joints) in a 28 year-old woman (Casey et al. 2013).

The case described is that of an actual patient who presented to the emergency department complaining of fatigue and joint pain. She gave no clinically significant medical history. She initially had pain in the right foot and ankle, which limited her ability to walk. These symptoms persisted for a few days, and then resolved, but then pain in her knees and hips developed, along with pain and swelling in her right elbow. She also reported intermittent low-grade fevers, and denied a recent cough or sore throat.

Although the range of possible causes of joint pain and fatigue is broad, the polyarticular, migratory nature of her joint pain helped narrow the possibilities. The

---

[4]For example, laboratory tests may be in error, or signs or symptoms may be evolving or misinterpreted.

differential diagnosis at this point included a viral syndrome, postinfectious or reactive arthritis, gonococcal infection with associated arthritis, and systemic rheumatic illnesses, including rheumatoid arthritis. Crystalline arthropathies like gout can be polyarticular, but this class was thought to be unlikely due to her young age and the presence of a systemic symptom (fatigue).

Pertinent findings on examination included a temperature of 38.3 °C (101.0 °F) and diffuse tenderness of the joints on palpation, but no swelling or limitation of motion was present. Cardiac examination revealed a tachycardia (increased heart rate), but no murmurs were detected. A chest radiograph showed no evidence of infection, and the white blood cell count was elevated. The fever and elevated white-cell count did not help to distinguish infectious causes from inflammatory causes. She was given a diagnosis of viral infection and discharged with a course of non-steroidal anti-inflammatory drugs.

She returned to the emergency department 5 days later with progressive shortness of breath. Her joint pain had diminished, but she still had symptoms in her right knee. The left knee was warm, and painful when flexed, but without swelling. She had a slight tachycardia with a new heart murmur consistent with mitral valve regurgitation. Signs and symptoms were consistent with congestive heart failure. Her temperature was 37.5 °C (99.5 °F), with normal blood pressure and respirations.

The combination of an apparently new regurgitant murmur and recent fever is suggestive of infective endocarditis, with the former being a major Duke criterion and the latter a minor Duke criterion for the diagnosis. The diagnosis of infective endocarditis is established when two major, one major plus three minor, or five minor Duke criteria are met. Obtaining blood cultures is thus important, since bacteremia (bacteria in the blood) with a causative organism is a major Duke criterion and would establish the diagnosis. The reported findings on physical examination did not include any of the stigmata of infective endocarditis, such as petechiae (small hemorrhages under the skin) or conjunctival hemorrhages.

Other causes of mitral regurgitation with heart failure were considered, but were believed to be unlikely for various reasons. For example, the most common cause of mitral regurgitation, mitral-valve prolapse, can lead to heart failure, but would not be expected to have such a fulminant course. Although rare in developed countries, it was suggested that acute rheumatic fever should also be considered.

She had never sought medical care before, and so, as noted previously, had no medical history. She was born in Brazil and moved to the northeastern United States 10 years before presentation. She was employed as a preschool teacher. She denied being sexually active, or using alcohol or intravenous drugs. She denied medication use or drug allergies.

The authors noted that the patient had no documented risk factors for infective endocarditis, such as human immunodeficiency virus (HIV) infection or end stage renal disease, but the fact that she was born in Brazil placed her at increased risk for rheumatic heart disease, which is a risk factor for both infective endocarditis and recurrent rheumatic fever. Rheumatic heart disease is a minor Duke criterion, and patients that have had acute rheumatic fever are at high risk for recurrent disease after subsequent group A streptococcal infections.

Acute rheumatic fever is a complication of pharyngeal infection with group A streptococcus, and symptoms usually develop 2–3 weeks after infection. However, subclinical episodes do occur. The absence of a history of sore throat in this patient does not reduce the likelihood of acute rheumatic fever, since most patients with the diagnosis of acute rheumatic fever do not recall a previous sore throat, even when there is serologic evidence of recent infection. A migratory arthritis can be one of the predominant symptoms of acute rheumatic fever.

Some patients with rheumatic heart disease recall having a low exercise tolerance in childhood. On further questioning, she recalled that when growing up in Brazil, she became short of breath easily, and was unable to play with the other children. At this point, with this history of shortness of breath in childhood, along with the combination of fever, mitral regurgitation, and migratory arthritis, acute rheumatic fever became the most likely diagnosis.

Acute rheumatic fever is diagnosed by meeting the Jones criteria and by obtaining evidence of a recent group A streptococcal infection. The major Jones criteria are migratory arthritis, carditis, chorea (irregular, involuntary muscular movements), erythema marginatum (a type of acute skin eruption), and subcutaneous nodules. The minor criteria are arthralgia, fever, elevated blood levels of acute phase reactants (proteins associated with stressful or inflammatory states), and a prolonged PR interval (an electrocardiographic finding). To satisfy the diagnostic criteria, either two major criteria or one major and at least two minor criteria must be met. This case met two major criteria (migratory arthritis and carditis) and one minor criterion (fever on initial presentation to the emergency department). To confirm the diagnosis, evidence of recent infection with group A streptococcus is required. Throat culture has a low diagnostic sensitivity, since the symptoms of acute rheumatic fever appear 2–3 weeks after the antecedent streptococcal infection, by which time throat cultures are negative in many patients. Evidence for group A streptococcal infection in this patient was obtained by detecting elevated titers of antistreptolysin and anti-DNase B antibodies in the blood stream. Evidence against infective endocarditis came from negative blood cultures, indicating the absence of bacteremia. Transthoracic and transesophageal echocardiograms demonstrated structural heart disease, with a combination of severe mitral regurgitation and stenosis (narrowing). In the vast majority of cases, mitral stenosis is caused by rheumatic disease.

The diagnosis of acute rheumatic fever in this patient was confirmed by meeting specified diagnostic criteria. However, it is conceded that there is still no single symptom, sign, or laboratory test that is pathognomonic[5] or diagnostic of acute rheumatic fever (Dajani et al. 1993, 302). When Jones first advanced his criteria for the diagnosis in 1944, he did so with the intention to reduce variation in diagnostic criteria, which differed widely by observer, and he contended that ". . . it would seem logical to make a positive diagnosis on rather strict criteria" (Jones 1944, 484). His original criteria were divided into "major manifestations" and "minor manifestations," the former offering ". . . the least likelihood of an improper diagnosis" (Jones

---

[5]A "pathognomonic" finding is one that can occur only with a single condition.

1944, 481). Carditis was included as a major manifestation because active carditis was found in all fatal cases of rheumatic fever, and arthralgia because migrating polyarthritis was generally considered the classic feature of rheumatic fever. He included chorea as a major manifestation because, based on his own reported data, about one half of all young rheumatic fever patients have chorea at some time, and about three-fourths of young patients with chorea in time develop other major manifestations of rheumatic fever. He lists five major manifestations in all, and states that a combination of them makes a diagnosis of rheumatic fever reasonably certain.

He advanced seven minor manifestations, but stated that even a combination of them may not be sufficient to make a certain diagnosis, but might be suggestive. He did suggest, however, that any single major manifestation with at least two of the minor manifestations would seem to place the diagnosis on "reasonably safe grounds" (Jones 1944, 483).

The Jones criteria have been periodically modified, revised, and updated by the American Heart Association, the most recent update being that presented by Dajani et al. (1993). Thus it appears that the criteria were originally based on a somewhat loose appreciation of the relative frequencies of the various manifestations of the condition as it evolves in patients. Exceptions to the Jones criteria have also been advanced (Dajani et al. 1993, 307).

It will be recalled that one of the initial diagnostic hypotheses considered in the case of rheumatic fever discussed above was infective endocarditis, but that diagnosis was essentially eliminated when the Duke criteria were not met. The Duke criteria have been relatively extensively evaluated in numerous studies in both Europe and the United States, and the sensitivity and specificity have been reported. Using the value of 80% for the sensitivity (Li et al. 2000) and 99% for the specificity (Hoen et al. 1996), the positive predictive value is 99%, as illustrated in Table 4.7:

Sensitivity = probability of meeting the DC when IE is present = .80
Specificity = probability of not meeting the DC when IE is absent = .99
Positive Predictive Value = probability of having IE when the DC are met = 80/
    81 = .99

It will be noted that the positive predictive value, the probability of disease (in this case infective endocarditis) given the evidence (in this case meeting the Duke criteria) is the left hand side of the familiar Bayes' equation.

Thus objective probabilities are available in the case of Duke criteria for the diagnosis of infective endocarditis, but are not available for the Jones criteria for diagnosing rheumatic fever. This is so because a "gold standard," pathological

Table 4.7 Hypothetical population with 80% sensitivity and 99% specificity

	IE present	IE absent	
DC met	80	1	81
DC not met	20	99	119
	100	100	200

DC Duke Criteria, *IE* Infective Endocarditis

confirmation, was used to define cases of infective endocarditis, against which the probability that various criteria could successfully predict presence or absence of disease was assessed. No such gold standard for rheumatic fever has yet been determined.

This application of Bayes' theorem raises the possibility that Bayesian confirmation theory may at least partially underlie the process of reasoning in differential diagnosis, and indeed it has been postulated that there exists a close parallelism between the implicit reasoning processes that physicians use to revise and refine diagnostic hypotheses with new information and the formal prescriptive process that calculates these revisions (Elstein and Schwarz 2002). Kassirer et al. (2010, 21–23) have discussed Bayesian analysis in the context of differential diagnosis, illustrated by a case of a man with renal insufficiency. The man's examination and a variety of laboratory studies had narrowed the diagnostic possibilities of the cause to five conditions: glomerulonephritis (GN), interstitial nephritis (IN), acute tubular necrosis (ATN), functional acute renal failure from dehydration (FARF), and atheromatous embolism (AE). The diagnostic importance of two physical findings, hypertension and livedo reticularis (a type of purplish skin discoloration), and two laboratory findings, sparse urine sediment and low complement (CH50), were assessed. Approximate conditional probabilities of the four factors were obtained from a literature survey, and the results of a Bayesian analysis of the prior probabilities, conditional probabilities, and calculated posterior probabilities are shown in Table 4.8:

According to the authors, several observations can be made. A "diagnosis" is in truth a *probability distribution* for a set of diagnostic possibilities, which in this case are the various causes of acute renal failure. Also, the estimate of the prior probabilities of any given diagnosis and the relation between the conditional probabilities substantially affect the posterior probabilities. For example, the prior probability of .29 for glomerulonephritis made it a serious candidate initially, and this was supported by the findings of hypertension and low complement. However, the rarity of a sparse urine sediment and livedo reticularis in this disorder rendered the posterior probability of .019, which is quite low. Also, atheromatous embolism was quite unlikely initially, with a prior probability of .01, but because the

Table 4.8 Bayesian analysis for acute renal failure

Disease	Prior Pr	Conditional probabilities				Posterior Pr
		Htn	Lr	Sparse sed	Low CH50	
GN	.29	.60	.05	.01	.40	.019
IN	.10	.10	.05	.15	.01	<.01
ATN	.40	.05	.05	.15	.01	<.01
FARF	.25	.01	.20	.95	.01	<.01
AE	.01	.80	.60	.95	.40	.977

Source: Adapted from Kassirer et al. (2010). p. 22, table 4.8
Pr probability, *Htn* hypertension, *Lr* livedo reticularis, *sed* sediment, *CH50* complement, *GN* glomerulonephritis, *IN* interstitial nephritis, *ATN* acute tubular necrosis, *FARF* functional acute renal failure from dehydration, *AE* atheromatous embolism

conditional probabilities overall of the factors considered was higher than in the other disorders, the posterior probability of .977 made the diagnosis of atheromatous embolism the most likely.

As the authors point out, several conditions are either necessary or highly desirable in order to apply Bayes' theorem. Definitions of all the diagnostic hypotheses under consideration should be unambiguous. Most desirable is that they be based on some "gold standard," that is, some generally accepted and relatively irrefutable evidence that a certain condition exists, such as in the pathological confirmation for the Duke criteria above. A histologic diagnosis is the most common type of such evidence, although in some cases biochemical or genetic markers may substitute. Disease attributes that form the basis of conditional probabilities may vary according to factors such as age or stage of disease; in addition, diseases are often unstable and evolve over time, and thus the probability of certain attributes may likewise vary over time. Thus, care must be taken when formulating attributes that will serve as conditional probabilities.

Other considerations include the lack of a need to list every possible diagnostic hypothesis separately; many may be combined into a "catchall hypothesis." In the above example, this might take the form of a category labeled "other etiologies of acute renal failure." Of course, all relevant diagnoses must be included or the correct diagnosis might never be made. In addition, each diagnostic hypothesis must be mutually exclusive of all the others under consideration, and each conditional probability used in a calculation must be independent of the others. Also noted is that certain diseases cannot be considered as being simply present or absent. Often stages of diseases have different manifestations, thus quantitative analyses must recognize that attributes may vary over time. Thus, particularly when simultaneously considering many findings or diagnostic hypotheses, mathematical predictive modeling methods such as logistic regression are often preferred.

Eddy and Clanton (1982) have also evaluated a potential role for Bayes' theorem in differential diagnosis. They note that, ideally, in order to select the most probable diagnosis, the physician would need to calculate and compare the probabilities of the diseases that could have caused the patient's signs and symptoms. The most direct method would be to use Bayes' theorem, however the obstacles to employing Bayes' theorem in the vast majority of cases where diagnoses are actually made seem almost insurmountable. These include the vast amount of information to be considered, the need to link signs and symptoms to disease even though medical knowledge is learned primarily by disease, and the need to be versatile calculating and otherwise using probabilities. For these and other reasons they believe that it is unlikely that the method of reasoning used by physicians to perform complicated diagnoses resembles the actual application of the Bayes' equation.

Eddy and Clanton state the diagnostic problem thusly: "When a clinician encounters a patient, the clinician faces a vast amount of information: the patient's lifelong personal and medical history; the patient's report of the current medical problem; and the results of numerous examinations, procedures, and tests. In addition to this information the clinician must have a tremendous amount of knowledge about health and disease." In spite of these barriers, however, they go on to note: "Somehow, seasoned clinicians are able to sort their way through the details, clear the confusion, and make the diagnosis" (1982, 1263).

Eddy and Clanton studied 50 case reports published as clinicopathologic conferences with the aim of elucidating the principles of reasoning employed by clinicians when forming, assessing, and confirming various diagnostic hypotheses. Their study suggested that "... six steps are used to arrive at a clinical diagnosis: aggregation of elementary findings, selection of a "pivot" (or pathognomonic finding), generation of a cause list, pruning of the cause list, selection of the diagnosis, and validation of the diagnosis" (1982, 1264). Elementary findings are single pieces of information about a case, for example cough with sputum production or a heart murmur. In a typical case, the discussant is presented with hundreds of elementary findings, which are combined to produce aggregates. Thus, some aggregate of findings may suggest a particular disease. The goal is to find an explanation or cause of a set of findings.

In very rare cases there is a pathognomonic finding. Lacking this, the discussant moves from a list of findings to a list of causes for those findings. Usually one or possibly two particularly salient findings are selected for focus, which the authors call the selection of a "pivot." After selecting the pivot, the possible diseases that could have caused the pivot are considered. Focusing on the pivot eliminates the need to focus on the probabilities of various diagnostic hypotheses: the discussant is not concerned with the probabilities of the diseases on the list, only with whether any could have caused the pivot. Pruning of the cause list begins by inspecting the diseases on the cause list one at a time. Implausible causes of the pivot are eliminated. Diseases with very low likelihoods are also eliminated, and if more than one disease remains, this pruned cause list becomes a tentative differential diagnosis of the case.

In pruning the cause list, although the discussant is searching for the most probable diagnosis, no probabilities are estimated. What is determined is whether the pattern of findings in the case could have been caused by any of the diseases on the list. As the authors note, this is a comparison rather than a calculation, and it uses knowledge of the characteristics of diseases instead of requiring estimation of the probabilities of a disease, given the findings (i.e., $P(D/e)$, the left hand side of Bayes' equation). Most discussants use the heuristic of comparing two diseases at a time, which is theoretically correct. If one always chooses the more likely of two diseases in these comparisons, the most probable disease will emerge. It allows selection of the most probable disease without the need to specifically estimate probabilities.

They also point out that these published cases are usually complicated, and that in clinical practice far more often most problems are much simpler and easier to solve, requiring little more than aggregation. More common are patients that present with the classical findings of one or two conditions, and the diagnosis is readily made. They do believe, however, that the case reconstructions used in published clinicopathologic conferences are in many ways similar to what goes on in actual clinical practice, as evidenced in conferences, rounds, consultations, and the like.

It will be observed that in some respects diagnostic hypotheses differ from therapeutic and etiologic hypotheses. For example, with the exception of the N of 1 study, the latter two types of hypotheses are concerned with groups of persons and, as we have seen, these groups can sometimes be quite large and number into the hundreds or even thousands. In contrast, diagnostic hypotheses and the N of 1 study are concerned with a single person. If the goal of science is to provide generalized, abstract knowledge, then how, it may be asked, or even why, should an account of evidence apply to both populations and individuals?

The answer, I believe, lies in some very basic considerations. The domain of the inductive, empirical sciences includes, at least in principle, all phenomena in the natural world, which of course would include individuals or groups. A single person suffering from a malady should be as open to scientific enquiry concerning the cause, for example, as would be a group of people suffering the same malady, as in an epidemic. On this view, what makes a hypothesis "scientific" is, at least in part, how evidence is gathered and used in the process of confirming or disconfirming the hypothesis. If the hypothesis is empirical, then in principle it should be open to scientific enquiry. And, as we have seen, the theoretical underpinning and reasoning process for diagnostic hypotheses is Bayesian (Elstein and Schwarz 2002; Kassirer et al. 2010, 21–23).[6]

As an example, consider the tea-tasting lady described by Mayo (1996, 154–160) and discussed earlier. The lady claimed to be able to tell by tasting whether milk was added to the cup before or after the tea was poured. This claim was susceptible to scientific investigation, it will be recalled, and an experiment was set up consisting of 100 tastings under carefully controlled conditions to eliminate bias. If she is just guessing, she should guess right about half the time. If her claim is true, she would be expected to do much better than just guessing. The binomial theorem was used to test the hypothesis. The fact that the hypothesis concerned only a single individual is irrelevant. This same reasoning, I believe, would apply to the numerous other examples in the philosophical literature on evidence in which hypotheses are concerned with a single person.

Another difference between diagnostic hypotheses and (again with the exception of N of 1 studies) therapeutic and etiologic hypotheses is that the latter two are not only studied in groups, but in addition the purpose for both is dissemination to the relevant expert community (through conferences, publications, etc.) and to become generalized, abstract scientific knowledge. Diagnoses are for the most part made in the course of the ordinary practice of medicine, and thus remain local, although some interesting or unusual cases will appear in the scientific medical literature as, for instance, a case report. Further, case reports bearing interesting similarities are sometimes assembled and reported together as a series. In this connection at least one commentator has made a distinction between "medical science" and "scientific medicine" (Miettinen 2001a). Thus for example RCTs and etiologic studies would, on this view, be classed as medical science, whereas medical diagnosis (and pre-sumably the N of 1 study) would be classed as examples of scientific medicine (Miettinen 2001b).

---

[6]Additional references include Gill et al. (2005), Willis et al. (2013), and Jain (2017).

The most important common factor that unites the three types of hypotheses is that all are designed to answer a question or solve a problem in clinical medicine, and the approach used must meet standards of scientific rigor that maximize the probability of achieving accurate results. For RCTs and etiologic studies this means following established procedures for their conduct and analysis, and for diagnostic hypotheses it means generating hypotheses from initial signs and symptoms, gathering of data to test hypotheses, and elimination of improbable hypotheses in the attempt to arrive at the correct diagnosis. The most common diagnoses are considered first, and data collection is hypothesis driven (Sox et al. 2007, 9–26). Misdiagnosis, the labeling of the patient with an incorrect diagnosis, is more likely to result if a sufficiently rigorous procedure is not followed (e.g., Berghmans and Schouten 2011).

For simplicity, in the present work I have included therapeutic, etiologic, and diagnostic hypotheses together under the rubric "clinical medical science" since all are probabilistic hypotheses in which evidence can be evaluated by statistical or probabilistic methods that are widely used and generally accepted by the scientific community.

References

Anderson, Kristin E., Thomas M. Mack, and Debra T. Silverman. 2006. Cancer of the pancreas. In *Cancer epidemiology and prevention*, ed. David Schottenfeld and Joseph F. Fraumeni Jr., 3rd ed., 721–762. Oxford: Oxford University Press.

Berghmans, Ron, and Harry C. Schouten. 2011. Sir Karl Popper, swans, and the general practitioner. *British Medical Journal* 343: d5469.

Casey, Jonathan D., Daniel H. Solomon, Thomas A. Gaziano, Amy Leigh Miller, and Joseph Loscalzo. 2013. A patient with migrating polyarthralgias. *New England Journal of Medicine* 369: 75–80.

Checkoway, Harvey, Neil E. Pearce, and Douglas J. Crawford-Brown. 1989. *Research methods in occupational epidemiology*. Oxford: Oxford University Press.

Clavel, Françoise, Ellen Benhamou, Ariane Auquier, Michèle Tarayre, and Robert Flamant. 1989. Coffee, alcohol, smoking and cancer of the pancreas: A case-control study. *International Journal of Cancer* 43: 17–21.

Cox, D.R. 1972. Regression models and life-tables. *Journal of the Royal Statistical Society, Series B* 34: 187–220.

Dajani, Adnan S., Elia Ayoub, Fredrick Z. Bierman, Alan L. Bisno, Floyd W. Denny, David T. Durak, Patricia Ferrieri, et al. 1993. Guidelines for the diagnosis of rheumatic fever: Jones criteria, updated 1992. *Circulation* 87: 302–307.

Daling, Janet R., and Karen J. Sherman. 1996. Cancers of the vulva and vagina. In *Cancer epidemiology and prevention*, ed. David Schottenfeld and Joseph F. Fraumeni Jr., 2nd ed., 1117–1129. Oxford: Oxford University Press.

Doll, Richard, and A. Bradford Hill. 1950. Smoking and carcinoma of the lung. Preliminary report. *British Medical Journal* 221: 739–748.

Doll, Richard, Richard Peto, Jillian Boreham, and Isabelle Sutherland. 2004. Mortality in relation to smoking: 50 years' observations on male British doctors. *British Medical Journal* 328: 1519–1528.

Eddy, David M., and Charles H. Clanton. 1982. The art of diagnosis. Solving the clinicopathological exercise. *New England Journal of Medicine* 306: 1263–1268.

Elstein, Arthur S., and Alan Schwarz. 2002. Clinical problem solving and diagnostic decision making: Selective review of the cognitive literature. *British Medical Journal* 324: 729–732.

Gill, Christopher J., Lora Sabin, and Christopher H. Schmid. 2005. Why clinicians are natural bayesians. *British Medical Journal* 330: 1080–1083.

Greenhalgh, Trisha. 2010. *How to read a paper: The basics of evidence-based medicine*. 4th ed. Chichester: Wiley.

Gutensohn, Nancy, Frederick P. Li, Ralph E. Johnson, and Philip Cole. 1975. Hodgkin's disease, tonsillectomy, and family size. *New England Journal of Medicine* 292: 22–25.

Guyatt, Gordon, Sackett David, D. Wayne Taylor, John Chong, Robin Roberts, and Stewart Pugsley. 1986. Determining optimal therapy – Randomized trials in individual patients. *New England Journal of Medicine* 314: 889–892.

Harvey, A. McGehee. 1994. Foreword. In *Differential diagnosis*, ed. Jeremiah A. Barondess, Charles C.J. Carpenter, and A. McGehee Harvey. Philadelphia: Lea and Febiger.

Herbst, Arthur L., Howard Ulfelder, and David Poskanzer. 1971. Adenocarcinoma of the vagina. Association of maternal stilbestrol therapy with tumor appearance in young women. *New England Journal of Medicine* 284: 878–881.

Hoen, Bruno, Isabelle Béguinot, Christian Rabaud, Roland Jaussaud, Christine Selton-Suty, Thierry May, and Philippe Canton. 1996. The Duke criteria for diagnosing infective endocarditis are specific: Analysis of 100 patients with acute fever or fever of unknown origin. *Clinical Infectious Diseases* 23: 298–302.

Hosmer, David W., and Stanley Lemeshow. 1989. *Applied logistic regression*. New York: Wiley.

International Agency for Research on Cancer. 2004. *Tobacco smoke and involuntary smoking*, IARC monographs on the evaluation of carcinogenic risks to humans. Vol. 83. Lyon: IARC.

Jain, Bimal. 2017. The scientific nature of diagnosis. *Diagnosis* 4: 17–19.

Jones, T. Duckett. 1944. The diagnosis of rheumatic fever. *Journal of the American Medical Association* 126: 481–484.

Kaplan, E.L., and Paul Meier. 1958. Nonparametric estimation from incomplete observations. *Journal of the American Statistical Association* 53: 457–481.

Kassirer, Jerome P., John B. Wong, and Richard I. Kopelman. 2010. *Learning clinical reasoning*. 2nd ed. Baltimore: Lippincott Williams and Wilkins.

Kelsey, Jennifer L., W. Douglas Thompson, and Alfred S. Evans. 1986. *Methods in observational epidemiology. Volume 10 of monographs in epidemiology and biostatistics*, General ed. Abraham M. Lilienfeld. Oxford: Oxford University Press.

Li, Jennifer S., Daniel J. Sexton, Nathan Mick, Richard Nettles, Vance G. Fowler Jr., Thomas Ryan, Thomas Bashore, and G. Ralph Corey. 2000. Proposed modifications to the Duke criteria for the diagnosis of infective endocarditis. *Clinical Infectious Diseases* 30: 633–638.

MacMahon, Brian, Stella Yen, Dimitrios Trichopoulos, Kenneth Warren, and George Nardi. 1981. Coffee and cancer of the pancreas. *New England Journal of Medicine* 304: 630–633.

Mantel, Nathan, and William Haenszel. 1959. Statistical aspects of the analysis of data from retrospective studies of disease. *Journal of the National Cancer Institute* 22: 719–748.

Mayo, Deborah G. 1996. *Error and the growth of experimental knowledge*. Chicago: University of Chicago Press.

Miettinen, Olli S. 2001a. The modern scientific physician: 2. Medical science versus scientific medicine. *Canadian Medical Association Journal* 165: 591–592.

———. 2001b. The modern scientific physician: 3. Scientific diagnosis. *Canadian Medical Association Journal* 165: 781–782.

Morris, J.N., J.A. Heady, P.A.B. Raffle, C.G. Roberts, and J.W. Parks. 1953. Coronary heart-disease and physical activity of work. *Lancet* 2: 1053–1057.

Mueller, Nancy E. 1996. Hodgkin's disease. In *Cancer epidemiology and prevention*, ed. David Schottenfeld and Joseph F. Fraumeni Jr., 2nd ed., 893–919. Oxford: Oxford University Press.

Mueller, Nancy, G. Marie Swanson, Chung-cheng Hsieh, and Philip Cole. 1987. Tonsillectomy and Hodgkin's disease: Results from companion population-based studies. *Journal of the National Cancer Institute* 78: 1–5.

Rothman, Kenneth J., Sander Greenland, and Timothy L. Lash. 2008. Case-control studies. In *Modern epidemiology*, ed. Kenneth J. Rothman, Sander Greenland, and Timothy L. Lash, 3rd ed., 111–127. Philadelphia: Lippincott Williams and Wilkins.

Slone, Dennis, Samuel Shapiro, Lynn Rosenberg, David W. Kaufman, Stuart C. Hartz, Allen C. Rossi, Paul D. Stolley, and Olli S. Miettinen. 1978. Relation of cigarette smoking to myocardial infarction in young women. *New England Journal of Medicine* 298: 1273–1276.

Sox, Harold C., Marshall A. Blott, Michael C. Higgins, and Keith I. Martin. 2007. *Medical decision making*. Philadelphia: American College of Physicians.

Straus, Sharon E., Glasziou Paul, W. Scott Richardson, and R. Brian Haynes. 2011. *Evidence-based medicine. How to practice and teach it*. 4th ed. Edinburgh: Churchill Livingstone.

Thomas, Patrick R.M., and Anne S. Lindblad. 1988. Adjuvant postoperative radiotherapy and chemotherapy in rectal carcinoma: A review of the gastrointestinal tumor study group experience. *Radiotherapy and Oncology* 13: 245–252.

Ueshima, Hirotsugu, Takashi Shimamoto, Minoru Iida, Masamitsu Konishi, Masato Tanigaki, Mitsunori Doi, Katsuhiko Tsujioka, et al. 1984. Alcohol intake and hypertension among urban and rural Japanese populations. *Journal of Chronic Diseases* 37: 585–592.

Willis, Brian H., Helen Beebee, and Daniel S. Lasserson. 2013. Philosophy of science and the diagnostic process. *Family Practice* 30: 501–505.

Wynder, Ernest L., and Evarts A. Graham. 1950. Tobacco smoking as a possible etiologic factor in bronchiogenic carcinoma. *Journal of the American Medical Association* 143: 329–336.

Chapter 5
A Weight of Evidence Account

Abstract In this chapter, I explicate the weight of evidence account. The weight of evidence accorded a hypothesis is highly dependent on the *accuracy* of the evidence, which is understood to mean a combination of validity and precision. Validity is concerned with biases and confounding, which can distort the measure of effect, whereas precision is often associated with the width of confidence intervals around the effect measure. Evidence considered to be of greater accuracy is accorded greater weight. The weight of evidence notion can be applied to single observations or studies, or groups of these. For therapeutic hypotheses, the randomized clinical trial affords the greatest degree of accuracy. Threats to validity in etiological studies are greater, sometimes resulting in lower weights being accorded. Maximum weight of evidence can be achieved for diagnostic hypotheses by meeting a "gold standard." The processes by which evidence is determined and quantified are considered, and the ways in which evidence is aggregated, evaluated, and applied to problems including determining best therapy or improving public health policymaking are discussed.

5.1 General

In this chapter I will explicate a notion of evidence that I believe makes sense of the way that evidence is gathered and used in clinical medical science. In doing so, and uncertain of the interests and backgrounds of readers, the presentation is somewhat detailed. To some, parts may possibly seem somewhat repetitive or overemphasized. Others may find parts that seem unnecessary or even trivial. I believe it is more important to err on the part of clarity and completeness rather than possibly leave questions in the minds of readers, hence the detail. In addition, this chapter contains further examples and discussion of confounding and bias in observational studies (which as I indicated earlier has received insufficient attention in the literature on evidence), and examples that I will use later in defending this account.

The account to be elucidated here is a variation of the notion of *weight of evidence*. The term "weight of evidence" in support of a hypothesis is roughly

J. A. Pinkston, *Evidence and Hypothesis in Clinical Medical Science*, Synthese Library 426, https://doi.org/10.1007/978-3-030-44270-5_5

equivalent to expressions like "strength of evidence," "amount of evidential support," "degree of confirmation," or "degree of corroboration."

The term *weight of evidence* has been used previously. For example, it was used by Good (1960) in an attempt to sharpen Popper's notion of the degree to which evidence corroborates a hypothesis. As Mayo (2005, 105) has noted, Popper's notion of severe testing can be construed as a simple comparative likelihood account where e is evidence for H if P(e/H) > P(e/H'). Good (1960) defines weight of evidence to be

$$W(H : E|G) = \log \{P(E|H \cdot G)/P(E|\sim H \cdot G)\}$$

Here H is a hypothesis, $\sim$H is its negation, E is the (proposition expressing the) evidence obtained from observations or experiments, and G is any assumed proposition (often omitted in formulae). The colon means "provided by" and the vertical stroke means "given." Good notes that when H and $\sim$H are simple statistical hypotheses, W is the logarithm of a simple likelihood ratio, and also that any real function of P(E|H) and P(E|$\sim$H), that, together with P(H), mathematically determines P(H|E), is a monotonic function of W(H: E). Good (1985) also provides several examples of other uses of the term *weight of evidence*.

The notion of weight of evidence that I will advance is different from Good's or other previous notions, as will become apparent. For example, precise numerical quantification is not required although often obtainable, and while weight of evidence may in some instances be expressed as a probability, this is not necessary or even possible in others. The notion is objective, and the observations or experiments that constitute evidence are principally based on data that have been objectively obtained. In modern clinical medical science, evidence for a given hypothesis is compiled and evaluated as to the extent of weight that is accorded the hypothesis. The notion can be applied either to a single observation or study, or to groups of these.

The weight of evidence account strives to be a comprehensive explanatory instrument for clinical medical science in the spirit of Glymour's notion of the aims of confirmation theory (1980, 63–64). As such, it seeks not only to describe and explain how medical scientists make judgments about the strength of evidence, but also to describe and explain the determinants of strength of evidence; that is, what makes evidence weak or strong, or of high or low quality. In addition, it can be construed as a causal theory in the sense that the hypotheses that we are considering are causal hypotheses. For therapeutic hypotheses, for example, in an ideal RCT, as Cartwright (2007) notes, if the assumptions of the test are met, a positive result implies the appropriate causal conclusion. And for studies of etiologic hypotheses and for diagnostic hypotheses, we are seeking the causes of disease or other adverse outcome and the causes of the patient's signs and symptoms, respectively.

The principal types of study used in clinical medical science include RCTs, meta-analyses of RCTs, non-randomized controlled clinical studies, cohort studies, case-control studies, cross-sectional studies, and case series. These are studies conducted on groups of subjects. The weight given to any individual study is directly proportional to the extent that the study is thought to be *accurate*. Accuracy here is to be understood to consist of a combination of *validity* and *precision*. Validity is concerned with systematic error; precision is concerned with random error or variation. Inferences from a study are accurate just to the extent that the study is valid and precise. The precision of a study can often be improved by increasing the number of study subjects, and sometimes a modification in study design can achieve a similar result. Validity is the more important of the two, and the greatest threats to validity come from biases and confounding. The notion of study accuracy as consisting of validity and precision is appreciated by researchers (e.g., Rothman et al. 2008, 128–129).

In experimental therapeutics, RCTs or meta-analyses of RCTs are thought to provide the greatest accuracy due to the theoretical minimization or elimination of potential sources of bias or confounding. In an RCT, eligible study subjects are randomly assigned to a therapeutic arm (e.g., an active drug) or to a control arm (e.g., a placebo), and all aspects of the study are controlled so that only the treatment variable (e.g., the active drug) is able to influence the treatment result. Assuming that the study was well-designed and conducted, and the data properly analyzed, possible biasing or confounding factors would be expected to be evenly distributed, thus minimizing or eliminating their influence. This is not to say that RCTs cannot be confounded, however. Rothman (1977) has considered the possibility of confounding in RCTs, and notes that methods for preventing or controlling confounding that are used in observational research may on occasion also have applicability for experimental research as well. Different therapeutic alternatives can of course be compared retrospectively by, for example, comparing different groups that have received different therapies at different times. However, such studies suffer from the possible influence of confounding and biases, which can be subtle. Thus, in general such studies are accorded less weight.

A randomization procedure is usually not available for studying etiologic hypotheses, thus bias and confounding are threats that are ever present and must be controlled to the extent possible. Sackett (1979, 51) reported that 35 biases had been cataloged that could occur in analytic research. However, the most important threats to validity come from selection and information biases, in addition to confounding. Although I have treated these issues to some extent earlier, they are of sufficient magnitude to warrant further discussion. Some examples are provided below.

5.2   Some Threats to Accuracy

5.2.1   Selection Bias

Selection bias results from methods of subject selection and factors that influence study participation. When the exposure-disease relationship is different for study participants than it would have been for all subjects eligible for inclusion in the study, then the estimate of effect may become distorted. One type of selection bias is *self-selection*.

An example of distortion due to self-selection occurred in the estimation of a leukemogenic effect secondary to radiation exposure from nuclear testing (Caldwell et al. 1980). In 1957 a nuclear device called "Smoky" was detonated at the Nevada test site. It was later estimated that over 3000 observers, mostly military troops, observed the event. In 1976 the Center for Disease Control received notice of a patient who developed leukemia, and who associated it with his exposure to the Smoky test. To test the hypothesis that an increase in leukemia cases was related to the Smoky test, an investigation was undertaken to identify the persons present at the test and their radiation exposure, and the subsequent development of leukemia. Nine cases of leukemia occurred among 3224 men who participated in military maneuvers during the explosion, but only 2459 members of the study cohort could be traced (76%). Extensive efforts to identify cohort members were undertaken, including widespread publicity in addition to record searching. The publicity produced responses from more than 3000 persons, of which 447 were confirmed to have been present at the Smoky test. Four cases of leukemia were identified among this group.

As Rothman et al. (2008, 134–135), note, a self-selection bias is present, consistent with previous studies that have shown that the reasons for self-referral may be associated with the outcome under study, and may be different from the remaining members (non-respondents) of the study population (Criqui et al. 1979). Of the 76% of the cohort that were traced, 82% were by the investigators and 18% were self-referred due to publicity. Thus, four leukemia cases were found among the $0.18 \times 0.76 = 14\%$ of cohort members that referred themselves, and four were found among the $0.82 \times 0.76 = 62\%$ of cohort members traced by the investigators. Thus, self-selection bias was present. How should we assess the expected number of leukemia cases in the 24% of the cohort that was untraceable? A different number of expected cases would result from the different assumptions that could be made. For example, if we assume that the 24% of the cohort that was not traced had a leukemia experience similar to the 62% traced by the investigators, we should expect 4 $(24/62) = 1.5$ or about one or two cases occurring in this 24%, resulting in 9 or 10 cases for the entire cohort. If instead we assume that the experience of the 24% was like that of the remaining 76% with known outcomes, then we should expect 8 $(24/76) = 2.5$ or about two or three cases in this group, resulting in 10 or 11 cases for the entire cohort.

Berkson (1946) described a different type of selection bias, called *Berkson's bias* or *Berksonian bias*. It is of particular concern in studies conducted among hospitalized patients, but may occur in other settings as well. It occurs because subject selection is affected by both exposure and disease, and specifically because it affects selection. For example, two diseases that are unassociated in the general population could be spuriously associated in a hospitalized population when both diseases affect the probability of hospital admission. So in a hypothetical case-control study in which the cases were hospitalized patients with disease x and controls were hospitalized patients with disease y, an exposure E that causes disease y would appear to be a risk factor for disease x. Berkson's bias and several other types of selection biases that may occur in etiologic studies are discussed by Hernán et al. (2004). Selection bias is also an important concern in therapeutic non-randomized controlled trials, in which selection factors may render the control group to be unsatisfactory to produce a valid estimate of a treatment effect (Green et al. 1997, 143–150).

5.2.2   Information Bias

Information biases result from a distortion in the estimation of effect that occurs from inaccurate measurement of the exposure or disease condition. This can result, for example, from the use of a measurement device (e.g., questionnaire or interview procedure) that does not measure what is intended, an inaccurate diagnostic procedure (for disease status), or an erroneous or incomplete data source. When situations such as these occur, a study subject may be misclassified as to exposure or disease status.

One type of information bias is *recall* bias. In etiologic studies using questionnaires, for example, information is often elicited from subjects that must recall relevant exposures. For example, in case-control studies of congenital malformations, one source of exposure information comes from interviews of mothers. The mothers of cases have recently given birth to a malformed baby, whereas control mothers have recently given birth to an apparently healthy baby. If the experience of having given birth to a malformed baby stimulated recall by mothers of exposures to drugs, trauma, or other potential factors related to malformations to a different extent than controls, this could be a source of bias (Rothman et al. 2008, 138). It has also been found that the amount of time between exposure and recall is an important indicator of the accuracy of recall. Differences in the time between exposure and recall, if different between cases and controls, could also produce bias (Klemetti and Saxén 1967).

5.2.3   Confounding

Confounding results from the distortion produced by one or more factors on the estimate of effect. Three necessary (but not sufficient or defining) characteristics have been ascribed to confounders (Rothman et al. 2008, 132–134): (1) A confounding factor must be an extraneous risk factor for the disease; (2) A confounding factor must be associated with the exposure under study in the source population (the population at risk from which the cases are derived); and, (3) A confounding factor must not be affected by the exposure or the disease. In particular, it cannot be an intermediate step in the causal path between the exposure and the disease.

A simple example of confounding is supplied by Gordis (1996, 185–186). Assume that a case-control study was performed to assess whether some exposure was associated with some disease, and that 100 cases of the disease and 100 controls without the disease were assembled without regard to possible bias or confounding, and that strategies like matching were not employed. Let us assume further that 30 of the cases and 18 of the controls were exposed. The exposure odds ratio is $(30 \div 70) \div (18 \div 82) = 1.95$. Thus, the odds of exposure among cases are nearly twice that of controls, which suggests that a relationship may exist between exposure and disease.

Suppose, however, that we consider whether the increased risk among cases might be due to a confounding factor. We will investigate whether age might be confounding the observed association. Thus we perform an age analysis as follows (Table 5.1):

We see that 80% of the controls are younger than 40 years compared with 50% of the cases. Thus, older age is associated with being a case (having the disease) and younger age with being a control (not having the disease). We next investigate whether age is related to exposure (Table 5.2):

We see that in studying the relationship of age to exposure in the total 200 subjects, that age is related to exposure status. Fifty percent of persons aged ≥ 40 years were exposed, whereas only 10% of those aged <40 years were exposed. Is the exposure-disease relationship causal or are we seeing the effect of a confounder, in this case, age? One approach to analyze this is to *stratify* on the basis of age, and carry out separate analyses for each age group (Table 5.3):

Thus, the only reason we had an odds ratio of 1.95 initially was because there was a difference in age distributions, and here age is a confounder.

Age is such a well – known potential confounder that it is routinely controlled for. Other potential confounders we have already considered in the examples thus far include cigarette smoking in studies of coffee drinking and pancreatic cancer, alcohol use in studies of cigarette smoking and lung cancer, and other diseases associated with myocardial infarction in studies of cigarette smoking and myocardial infarction in healthy young women. Control of confounding can be done before a

Table 5.1 Hypothetical example of confounding in an unmatched case-control study: II. Distribution of cases and controls by age

Age (year)	Cases	Controls
<40	50	80
≥ 40	50	20
Total	100	100

Source: Gordis (1996). p. 186, table 14-3

Table 5.2 Hypothetical example of confounding in an unmatched case-control study: III. Relationship of exposure to age

Age (year)	Total	Exposed	Not exposed	% Exposed
<40	130	13	117	10
≥ 40	70	35	35	50

Source: Gordis (1996). p. 186, table 14-4

Table 5.3 Hypothetical example of confounding in an unmatched case-control study: IV. Calculations of odds ratios after stratifying by age

Age (year)	Exposed	Cases	Controls		Odds ratio
<40	Yes	5	8		
	No	45	72	}	$\frac{5 \times 72}{45 \times 8} = \frac{360}{360} = 1.0$
	Total	50	80		
≥40	Yes	25	10		
	No	25	10	}	$\frac{25 \times 10}{25 \times 10} = \frac{250}{250} = 1.0$
	Total	50	20		

Source: Gordis (1996). p. 186, table 14-5

study is launched, for example, by age – matching cases and controls to control confounding by age, or after a study is completed, by, for example, stratification as in the above example. To control for the effects of a potential confounding factor, it must be *known* that the factor is potentially confounding, and investigators must also have accurate information about the presence or absence of the factor in the study population. Lack of such information is a major source of potential inaccuracy, especially in etiologic research.

5.3 How Is the Weight of Evidence Determined?

Modern computer technology and access to the Internet have allowed a potentially large amount of evidence to be compiled and evaluated for nearly all of the types of hypotheses encountered in clinical medical science. For hypotheses that have undergone extensive testing, such as those therapeutic and etiologic hypotheses believed by investigators to be the most important, the opportunity has arisen for expert panels or committees to be tasked with reviewing the available evidence. This has resulted in the publication of meta-analyses, systematic reviews, guidelines, and other reports that evaluate the evidence with the aim of providing a scientific justification for therapeutic decisions, as well as for health promotion and disease prevention efforts generally.

For therapeutic hypotheses, systematic reviews, like those done by Cochrane, and guidelines, such as those published by the National Comprehensive Cancer Network (NCCN), provide examples of the way that the weight of evidence is evaluated and established. For etiologic hypotheses, examples would include reports issued by organizations like the IARC and the U.S. Department of Health and Human Services. For diagnostic hypotheses, gold standards represent an example of maximum weight of evidence, such as biopsy confirmation for the diagnosis of breast cancer. Another example of maximum weight of evidence is meeting diagnostic criteria, such as satisfying the Duke criteria for the diagnosis of infective endocarditis.

In some cases the weight of evidence supporting a hypothesis may have accrued from a long history of success in clinical medicine. Examples would include appendectomy for acute appendicitis and screening by cervical cytological examination (the Papanicolaou test, or "Pap smear") for early diagnosis of cervical cancer. Evidential support for hypotheses such as these did not come from having been tested in clinical trials. In the modern era, however, efforts to improve on such well-established approaches most probably would require formal testing of some kind.

As I have indicated, the weight of evidence account can apply to a single observation or study, or to groups of these, and the same is true of accuracy. Several observations, such as results of physical examination or laboratory results in a case of differential diagnosis, are accurate just to the extent that the individual elements are accurate. For example, modern automation of many laboratory tests has resulted in a high degree of accuracy due to increased precision when compared with earlier methods. Variation among observers in the interpretation of physical findings or assessment of symptoms may occur, and increased precision can be enhanced through training and experience. This inter-observer variability may apply to any of the myriad encounters that involve interpretation by clinicians, including radiographic or pathological examination. This occurs frequently with diagnostic hypotheses, and instances of deference to "expert opinion" in the assessment of evidence could include, for example, the obtaining by a general hospital pathologist of a consultation with an expert academic pathologist in a case of difficult pathological diagnosis, or deferring to the opinion of a cardiologist by a general practitioner in the interpretation of heart sounds heard through a stethoscope. Insufficient accuracy could lead to misdiagnosis, in which flawed data lead to the erroneous confirmation of an alternative, incorrect diagnostic hypothesis. Thus, validity and precision considerations exist for diagnostic hypotheses as well as therapeutic and etiologic hypotheses.

The weight of evidence for a diagnostic hypothesis is just that which has accrued short of meeting a gold standard or diagnostic criteria (which would provide maximum weight of evidence). This is the situation, for example, that exists with a working diagnosis. When various diagnostic hypotheses are being considered, the situation is similar. Recalling the example of renal insufficiency and the use of Bayes' theorem in Table 4.8, discussed by Kassirer et al. (2010), in which several competing diagnostic hypotheses were evaluated, various observations and tests led to the hypothesis that atheromatous embolism was most likely the correct diagnosis since its posterior probability was .977. The gold standard for this diagnosis is a kidney biopsy, which would provide maximum confirmation. But with a posterior probability as high as .977, a biopsy might be avoided since biopsy confirmation can carry significant risk, particularly in the central nervous system or other internal organs. A gold standard should not necessarily be construed to mean that $\Pr = 1$, however, since it is possible for the application of the gold standard in any particular case to be in error.

Thus in these diagnostic hypothesis cases where Bayes' theorem could be used, the hypothesis with the highest probability has the support of the greatest weight of evidence. In cases where two hypotheses are being compared, which, according to

Eddy and Clanton (1982) represents the majority of cases of differential diagnosis, the hypothesis thought to have the highest probability has the support of the greatest weight of evidence. The weight of evidence account also easily accommodates any hypothesis involving a simple probability calculation. So if I own 800 tickets in a fair 1000 ticket lottery, for example, the probability that I will win is .800, and this would likewise represent the weight of evidence that I will win.

In the case of therapeutic hypotheses where several studies comprise the evidence, the weight of evidence account can be illuminated by the example of Cochrane Systematic Reviews.

5.3.1 Cochrane Systematic Reviews

According to its website (Cochrane 2019), *Cochrane* is an organization that exists so that healthcare decisions get better, and organizes its work around four goals: producing evidence, making the evidence accessible, advocating for evidence, and building an effective and sustainable organization. In producing evidence, their stated goal is: "To produce high-quality, relevant, up-to-date systematic reviews and other synthesized research evidence to inform health decision making." Cochrane has 13,000 members and over 50,000 supporters from more than 130 countries. Volunteers and contributors include researchers, health professionals, patients, and others. Cochrane has produced over 7500 Cochrane Systematic Reviews to date, and they are periodically updated as new information becomes available.

An example of a Cochrane Systematic Review is that assessing the value of adding radiotherapy to chemotherapy for early stage adult Hodgkin lymphoma (Blank et al. 2017). Combined modality therapy consisting of chemotherapy followed by localized radiotherapy for early stage (stages I and II) Hodgkin lymphoma is considered standard therapy, but because of long term adverse effects, including secondary malignancies, some have questioned the need for the addition of radiotherapy, suggesting that chemotherapy alone may be adequate treatment. A systematic review with meta-analysis of RCTs in which chemotherapy alone was compared with combined chemotherapy and radiotherapy in adult patients with early stage Hodgkin lymphoma was performed, with response rate, progression-free survival, and overall survival being the parameters of analysis. The method used for the accrual of the relevant studies was a search of databases such as the MEDLINE and CENTRAL databases, as well as conference proceedings, from January 1980 to December 2016 for RCTs in which chemotherapy alone was compared with combined modality therapy.

Two authors independently screened the titles and abstracts of potentially relevant RCTs that compared chemotherapy alone with chemotherapy combined with radiotherapy. The chemotherapy regimen had to be identical in both arms. The study was restricted to clinical stages I and II, according to predefined criteria. Exclusion criteria included trials with more than 20% of patients in more advanced stages, and trials of children with Hodgkin lymphoma.

Studies that met inclusion criteria from screening titles and abstracts were retrieved as full-text publications for detailed evaluation. The systematic review investigators contacted study authors to obtain missing information.

A total of 5518 potentially relevant records were identified through database searching. Sixty-four publications were retrieved for more detailed evaluation, and finally seven RCTs were used in the systematic review involving 2564 patients. These seven trials (24 publications) were included in both a qualitative synthesis and a quantitative synthesis (meta-analysis).

An assessment of the risk of bias in the included studies was made independently by two authors in order to comply with the Cochrane Handbook for Systematic Reviews of Interventions (Higgins and Green 2011). The following areas were examined (Blank et al. 2017, 11):

1. Was the allocation sequence adequately generated?
2. Was allocation adequately concealed?
3. Was knowledge of allocated intervention adequately prevented during the trial from patients, personnel, and outcome assessors (i.e., adequate "blinding")?
4. Were incomplete outcome data adequately addressed?
5. Are reports of the trial free of suggestion of selective outcome reporting?
6. Was the trial apparently free of other sources of bias (e.g. similarity of patients' characteristics at baseline)?

An example of one of the analyses that were performed in this review is that of five RCTs involving 1388 patients that compared the combination of chemotherapy alone and chemotherapy plus radiotherapy with the same number of chemotherapy cycles in both arms. The statistic used to report results was the hazard ratio (HR), which was 0.42 for progression free survival (95% C.I. 0.25–0.72; p = 0.001), and 0.48 for overall survival (95% C.I. 0.22–1.06; p = 0.07) for patients receiving radiotherapy plus chemotherapy compared with those receiving chemotherapy alone. The authors considered their evidence to be of moderate quality. They concluded that the addition of radiotherapy to chemotherapy improves progression free survival but probably has little or no effect on overall survival in early stage adult Hodgkin lymphoma. In the weight of evidence account, the weight of evidence for an improvement in progression free survival would be moderately strong, and the weight of evidence for an improvement in overall survival would be weak to none.

It is worthwhile noting here, however, that although meta-analysis as a method of deriving estimates of effect from the aggregation of data from several individual studies is widely used, it has sometimes been considered to be a controversial analytic method. There have, for example, been instances where different meta-analyses of the same evidence have produced different results.[1]

For etiologic studies where several studies comprise the evidence, an example that illuminates the weight of evidence account is provided by IARC reviews.

---

[1]For example, see Stegenga (2011). I will have more to say on meta-analysis later.

5.3.2 IARC Reviews

The IARC is the specialized cancer agency of the World Health Organization. A stated objective is to promote international collaboration in cancer research. The Agency is interdisciplinary and brings together skills in epidemiology, laboratory sciences, and biostatistics with the aim of identifying the causes of cancer. International expert working groups are formed to evaluate the carcinogenicity of specific exposures. The IARC is not directly involved in evaluating research on cancer care or therapeutics.

In the last chapter, we considered two IARC reviews. The 2004 review was concerned with the association between cigarette smoking and pancreatic cancer, in which it was concluded that the weight of evidence strongly supported an association. The 1991 review was on a possible association between coffee drinking and pancreatic cancer, in which it was concluded that studies on the whole were suggestive of a weak association between high levels of coffee drinking and pancreatic cancer, but that bias or confounding could account for the association. On their review of studies published since the IARC review, it will be recalled that Anderson et al. (2006, 741) noted that 11 case-control studies found null or non-significant associations between coffee consumption and pancreatic cancer, and that five cohort studies also found null associations. However, they also note that at least five studies *did* show statistically significant associations. They concluded that the possibility of an association between coffee drinking and pancreatic cancer seems unlikely and that reports of a significant association result from residual confounding from cigarette smoking, and possibly other sources of confounding or bias.

Thus, the weight of evidence is strong that cigarette smoking causes pancreatic cancer, but the weight of evidence does not support the hypothesis that coffee drinking is a cause of this disease.

5.4 How Is the Weight of Evidence Quantified?

For diagnostic hypotheses where sufficient information is available for an analysis using Bayes' theorem, which is rarely the case in practice, the probabilities that are produced can be directly translated into the weight of evidence account, as previously noted. The higher the probability of a hypothesis, the greater the weight of evidence accorded to it. In the more usual case, where diagnoses that have what are considered low probabilities are discounted, the one that is considered to have the highest probability becomes the working diagnosis. In any given case of differential diagnosis, more than one working diagnosis may appear as a disease evolves in its manifestations. Working diagnoses are not normally considered in precise probabilistic terms, but it would be reasonable to consider a working diagnosis to have a probability somewhere near .5 or greater, due to the fact that it is considered to have the highest probability among the alternatives being considered based on the evidence at hand. Should a diagnostic hypothesis meet a gold standard or criteria for the diagnosis of a specific disease, it would constitute maximum weight of evidence, as noted above. Also, it will be recalled that a patient may have more than one disease contemporaneously, thus a hypothesis meeting a gold standard or specific diagnostic criteria does not necessarily exclude other diagnoses.

The notion of probability being employed here for diagnostic hypotheses is frequency, and thus is objective. In principle, the probability that, given some group of signs and symptoms, a patient has a particular disease can be calculated, as was done in Chap. 4 with the case of renal insufficiency (Kassirer et al. 2010), and illustrated in Table 4.8. As Eddy and Clanton (1982) note, this is unwieldy and rarely done in practice; therefore signs and symptoms are aggregated, and attention is focused on certain findings, which may result in a certain salient finding that they refer to as a "pivot" that helps narrow the differential diagnosis.[2] When the differential diagnosis is sufficiently narrow, gold standard evidence is usually sought, particularly if the disease is serious, as was the case of the woman with acute rheumatic fever (Casey et al. 2013) discussed in Chap. 4, where the Jones criteria were met.

The medical literature is vast, and clinicians cannot master it all. However, for the common diseases, of which there are many, clinicians subsume a great quantity of information during primary training and subsequent educational activities, and a great deal of attention is focused on diseases and the signs and symptoms associated with them. Quantitative data are available and studied, but what is usually retained is a qualitative notion of the probability of some disease given the signs and symptoms that are present. In difficult cases quantitative data may be sought from the medical literature, as was done in the urinary insufficiency case (Kassirer et al. 2010).

Consider a hypothetical example:

A 75 year-old man presents to the urology clinic complaining of difficulty urinating and back pain. Evaluation discloses a prostate gland that is enlarged, firm, and nodular; a bone scan and x-rays reveal multiple blastic metastases in bones; and, his prostate-specific antigen (PSA) level is 1350 (normal <4.0).

This combination of signs and symptoms is practically diagnostic of prostate cancer, but why is this thought to be so? A Bayes' theorem analysis could be employed to obtain more precise quantitative data, and given the conditional probabilities that would no doubt result in a probability calculation for prostate cancer close to 1. But the reasoning is the same: It is a qualitative estimate of an objectively quantifiable probability estimate.

Thus when an assessment of the following sort is made, "These signs and symptoms point to disease x," or, (given these signs and symptoms), "Disease x is more probable than disease y," arguably a qualitative estimate of a quantifiable probability estimate is being made. It is based on reasoning using Bayes' theorem,

---

[2] A good example of a pivot, I would argue, is the finding of gastric folds in the patient with Ménétrier's disease (Lalazar et al. 2014) discussed in Chap. 3.

and it is a relative frequency concept of probability because in these cases, the use of Bayes' theorem is given a relative frequency interpretation.

For etiologic hypotheses, the relationship between an exposure and a disease is usually quantified according to measures of effect or association (Greenland et al. 2008). These include odds ratios, rate ratios, difference measures, and the like. As an example, consider the odds ratio in case-control studies. It is a measure of the relative effect of exposure on disease occurrence. The higher the odds ratio, the greater is the effect. Thus, *ceteris paribus*, a study resulting in an odds ratio = 3 would provide greater weight of evidence of a relative effect than would a study where the odds ratio = 1.5. The same general principle holds for the other measures.

Sometimes the term "relative risk" is used to quantify the measure of effect or association. An example of its use was provided in Table 4.3 and in the discussion of the risk of myocardial infarction among healthy women in relation to smoking status (Slone et al. 1978). It is a more general term and is based on odds ratios, rate ratios, and the like, and can also be used as a measure of the weight of evidence. Recall, for example, that I argued that the weight of evidence was very strong that cigarette smoking causes lung cancer based on a large body of evidence, which was illustrated by the British doctor study (Doll et al. 2004) and the two case-control studies (Doll and Hill 1950; Wynder and Graham 1950), where the p-values were very low. In a recent review of tobacco-related health conditions, estimates of the relative risks of mortality associated with tobacco use were presented. For current male smokers, for example, the relative risks were 23.3, 14.6, and 2.3 for cancers of the lung, larynx, and pancreas, respectively (Thun and Henley 2006, 219). Thus, relatively imprecise terms like "strong" or "moderate" in describing the weight of evidence can often be given more precise numerical quantification.

Relative risks substantially higher than the null value of 1.0 usually would be statistically significant at the conventional $p < .05$ level. For example, consider the odds ratio = 1.95 in the hypothetical example of confounding by Gordis (1996, 185–186) discussed earlier: This result is statistically significant (chi-square = 3.947; $p = 0.047$). Yet the observed effect was shown to be entirely erroneous, and was explained by confounding due to age. Thus, in etiologic studies in particular, we are reminded that the totality of the available evidence must be used to arrive at an accurate assessment of the weight of evidence for any given hypothesis, since etiologic studies are more subject to potential inaccuracy. This was amply demonstrated in the studies of, for example, cigarette smoking and lung cancer, where the relative effect of smoking was consistently very strong, and in the studies of coffee drinking and cancer of the pancreas, where it was concluded that it is doubtful that coffee drinking has any effect at all, notwithstanding that some studies showed a statistically significant association.

For therapeutic hypotheses, RCTs and the systematic reviews and meta-analyses based on them provide the greatest weight of evidence. Although cohort, case-control, and perhaps other study designs sometimes are used to evaluate therapeutic hypotheses, the relative lack of control over the many variables present renders them potentially less accurate. Where a group of RCTs comprise the evidence, the overall accuracy of the group is a function of the accuracy of the individual studies that comprise the group. Thus, for example, in the Cochrane systematic review cited above of the five RCTs that studied the value of adding radiotherapy to chemotherapy in early Hodgkin lymphoma (Blank et al. 2017), it will be recalled that the weight of evidence was considered moderately strong for an improvement in progression-free survival, and weak to none for an improvement in overall survival. In general, a larger number of RCTs studying the same therapy will provide the basis for a more accurate assessment of that therapy than would a smaller number.

The point estimates and C.I.s provide objective measures of therapeutic effect, and these in turn are related to the strength of the effect and the calculated p-values. The lower the p-value, the stronger is the effect. This is based on the notion that the stronger the effect, the less probable it is that the effect can reasonably be considered to be due to chance. Validity is relatively assured through the study design and randomization. Thus, for example, if two RCTs studying the same hypothesis differed *only* by the width of the confidence intervals around the same point estimate of effect, the RCT with the narrower C.I.s would be the more precise, and hence more accurate in my account. And, since it would be more accurate, it would therefore carry more weight.

Meta-analysis is an important tool for quantifying an overall measure of effect by aggregating the measures of effect and C.I.s from several studies of the same hypothesis. This was done in the Cochrane Systematic Review of the RCTs that compared chemotherapy alone with chemotherapy plus radiotherapy for early stage Hodgkin lymphoma described above (Blank et al. 2017). The reported HRs were less than 1.0 (null value) because they reflected the reduction in the risk to patients receiving chemotherapy plus radiotherapy compared with those that received chemotherapy alone. Meta-analyses are especially useful for groups of RCTs that are relatively homogeneous. Overall assessments of effect in systematic reviews that do not employ meta-analysis are by necessity usually not as precisely quantified.

Meta-analysis is less frequently employed in etiologic research due to the greater variability among studies and the greater threats from biases and confounding (Greenland and O'Rourke 2008). Data may be obtained differently in different studies, for example, by face-to-face interviews, mail-in questionnaires, or from record reviews. More generally, different studies may employ quite different protocols to collect, analyze, and report data. For example, Poole et al. (2006) reviewed the evidence linking socioeconomic status to childhood leukemia, and found such a large variation in the definition and measurement of socioeconomic status that they could only justify qualitative contrasts among the study results. Sometimes a review will simply list a range for the observed effect measures, as we saw in Chap. 4 was done with the studies of a possible relationship between previous tonsillectomy and Hodgkin lymphoma: In those studies, it will be recalled, the relative risks ranged from 0.7 to 3.6 (Mueller 1996, 912). Most often, it seems, the overall evidence for an etiologic hypothesis is expressed as none (e.g., coffee drinking and pancreatic cancer), weak, moderate, strong, or some variation of these, such as very strong (e.g., cigarette smoking and lung cancer).

However, when the relationship between exposure and disease is strong, even in etiologic studies a meta-analysis may sometimes be fruitfully undertaken. An example is the meta-analysis by Cannegieter et al. (1994) that evaluated the use of anticoagulant therapy on the risk of thromboembolism in patients that had undergone cardiac valve replacement. A total of 46 reports met inclusion criteria. The ratio of the incidence rates of thromboembolism in patients that did not receive anticoagulation was 5.6 (95% C.I. 4.2–7.5) compared with patients that received the anticoagulant warfarin. Thus the relative risk of thromboembolism associated with warfarin was 0.18 (95% C.I. 0.13–0.24). This large effect was considered high quality evidence (Guyatt et al. 2008, 997). This would be considered strong evidence in the weight of evidence account.

5.5 The Importance of Accuracy

Accuracy is an important factor on which the weight of evidence account is based. As I have indicated, the notion applies both to individual observations and studies as well as groups of these, and the accuracy of the groups is dependent on the individual observations or studies that comprise them. Although I have mentioned accuracy earlier, it is sufficiently important to merit further discussion.

Accuracy is, as I have noted, comprised by validity and precision. Observations and studies are accurate just to the extent that they are valid and precise. The most important threats to accuracy come from threats to validity, which most often result from biases or confounding.

Observations in the weight of evidence account are to be construed broadly. For diagnostic hypotheses, for example, they include (but are not limited to) findings on physical examination such as the observation of scleral pallor in a patient with anemia or the palpation of an abnormal lump in the breast, laboratory results, radiographic study results, and results of other diagnostic methods like electrocardiography. For etiologic studies they include items such as answers on a questionnaire, death certificate diagnoses, and the like; more generally, they are the individual data elements that comprise the large amount of information collected on exposure and disease status, and any factors that might be relevant to the exposure-disease relationship under investigation, particularly factors that might produce bias or confounding. For therapeutic hypotheses, similarly, a large number of factors are recorded in methods like RCTs (Green et al. 1997).

Data that are sufficiently inaccurate may fail to produce evidence for a true hypothesis, or provide putative evidence for a false hypothesis. In the latter case, it is not evidence at all. Following are some examples that help illuminate these concepts.

Consider the case of the young woman with migrating polyarthralgias discussed in Chap. 4 (Casey et al. 2013). Early in her diagnostic evaluation, infective endocarditis was considered in the differential diagnosis due to the detection of an apparently new regurgitant heart murmur. A new regurgitant heart murmur is a major Duke criterion for the diagnosis of this disease: The diagnosis is established when two major, one major plus three minor, or five minor criteria are met. Thus it was important to obtain blood cultures, since bacteremia with a causative organism is a major Duke criterion and would establish the diagnosis. Blood cultures were obtained, which were negative. It will be recalled that further investigation led to meeting the Jones criteria for acute rheumatic fever, which was the final diagnosis (presumed true hypothesis).

Suppose, however, that through faulty technique a blood culture became contaminated with a causative organism; suppose further that the standard procedure of obtaining several separate blood cultures was not followed and it was concluded that the patient indeed had bacteremia with the organism. This would be an example of a biased, hence invalid and inaccurate observation (putative bacteremia when it is not present) leading to the weight of evidence falsely being regarded as supporting the false hypothesis that the patient had infective endocarditis.

Other examples of inaccurate observations leading to false hypotheses would be a clinician mistaking the rash of measles for a rash secondary to an allergic reaction, or a pathologist misinterpreting a lymph node biopsy as lymphoma when in fact it is reactive hyperplasia, a benign condition. In a patient with a variety of signs and symptoms in which several diagnostic hypotheses are being considered in which sufficient information is available for a Bayesian analysis (rare in practice), the observations that constitute the data for the conditional probabilities, if sufficiently inaccurate, could significantly distort the posterior probabilities. This can be seen in the case of the man with renal insufficiency discussed in Chap. 4 (Kassirer et al. 2010) and pertinent data tabulated in Table 4.8, where inaccuracy in one or more of the four factors for which we have observations with conditional probabilities obviously could lead to inaccurate posterior probabilities.

Similar concerns with accuracy apply to etiologic studies. An etiologic study typically utilizes numerous observations that are organized in such a way as to assess whether one or more factors are related to the occurrence of disease or other adverse health outcome. If systematic error in the recording of one or more observations is present, for example the systematic misclassification of exposure or disease status in the individuals in the study, then bias may occur: The aggregate effect of the inaccurate data may lead to the study results themselves being inaccurately interpreted (e.g., as providing no evidence for a true hypothesis or falsely providing putative evidence for a false hypothesis).

Consider for example the case-control study by MacMahon et al. (1981) in which an elevated risk of pancreatic cancer was reported in coffee drinkers, after putative control for the known risk factor of cigarette smoking. Information on personal use factors such as cigarettes, coffee, or tea must be elicited directly from the subjects themselves, or sometimes from surrogates such as spouses or family members. Whatever the method, whether it be telephone interview, questionnaire, or some other tool, it is possible that the method itself may result in error such as bias. The exposure status of subjects may be systematically misclassified. It will be recalled that multiple other studies of the putative association between coffee consumption and pancreatic cancer failed to find such an association, notwithstanding that a very few did report an association. The few positive studies were later regarded as biased, probably as a result of residual confounding by cigarette smoking.

A more rigorous method such as an RCT can expose inaccuracy in prior observational research. Such a case is provided by the Nurses Health Study, a large cohort of nurses followed for health outcomes. For example, in a 10-year follow-up study of this group it was reported that current estrogen use in postmenopausal women was associated with a reduction in the incidence of coronary heart disease and in mortality from cardiovascular disease (Stampfer et al. 1991). These findings were later contradicted. For example, a subsequent RCT found that estrogen use in this group not only was not beneficial, but may actually increase the risk of coronary heart disease (Manson et al. 2003). As Mayo notes, the earlier observational studies were probably confounded, since women using the estrogens were found to have characteristics such as better health and education that are separately correlated with the beneficial outcomes (Mayo 2005, 97).

In cases such as these where observational studies are contradicted by more rigorous methods such as RCTs, inaccuracy due to unappreciated biases or confounding can be detected and evaluated. Sources of such inaccuracy could include, among others, failure to appreciate a factor as a risk factor and not collect information on it for the purpose of control, inaccurate recording of risk factor information, or systematic misclassification of exposure or disease status.

The RCT, and meta-analyses of RCTs, are widely regarded as providing the best (most accurate in the weight of evidence account) evidence for the types of hypotheses that can be studied by their use, which includes most therapeutic hypotheses. Nevertheless, they can produce inaccurate results. As Rothman (1977) has noted, RCTs can be confounded even when done properly, since randomization can fail to evenly distribute risk factors and there can be confounding due to chance. And not all RCTs have been carried out with strict adherence to method, resulting in poor RCTs with inaccurate results. In questionable cases, several RCTs may be required to provide needed evidence for any given hypothesis.

Meta-analyses may also be flawed and produce inaccurate estimates of effect. For example, Stegenga (2011) has argued that meta-analysis, even of RCTs, falls short of being the *platinum standard of evidence* that many regard it as being. For one, it may not adequately constrain the intersubjective assessments of hypotheses, because the numerous decisions to be made in designing and performing meta-analysis require personal judgment and expertise, which in turn allows personal biases to influence the outcome. This he suggests may partially account for cases in which multiple meta-analyses of the same evidence can reach contradictory conclusions. He also criticizes meta-analysis for relying on a narrow range of evidential diversity, suggesting that perhaps a broader method of amalgamating evidence for a hypothesis may be, at least in some cases, preferable.

It has also been alleged that there exists an overabundance of unnecessary, misleading and conflicted systematic reviews and meta-analyses that are suboptimal and serve as marketing tools or easily published units instead of promoting EBM or health care. One suggested policy approach is that "The publication of systematic reviews and meta-analyses should be realigned to remove biases and vested interests and to integrate them better with the primary production of evidence" (Ioannidis 2016, 485).

A recent review of meta-analysis and research synthesis outlines the significant advantages of properly performed meta-analyses, while acknowledging the problems that are encountered in their use (Gurevitch et al. 2018). The need for research synthesis will only increase with the advent of "big data" and artificial intelligence, and efforts to improve education of researchers in the techniques of proper research synthesis, including meta-analysis, are expected to increase.

Thus, accuracy is important in the weight of evidence account since the weight of evidence for any hypothesis is dependent on the accuracy of the observations or studies that constitute the evidence. Relative accuracy is the relative absence of error – both systematic and random. Understanding the sources of error and enabling efforts to eliminate them lie at the heart of understanding and improving confirmation in clinical medical science.

References

Anderson, Kristin E., Thomas M. Mack, and Debra T. Silverman. 2006. Cancer of the pancreas. In *Cancer epidemiology and prevention*, ed. David Schottenfeld and Joseph F. Fraumeni Jr., 3rd ed., 721–762. Oxford: Oxford University Press.

Berkson, Joseph. 1946. Limitations of the application of fourfold table analysis to hospital data. *Biometrics Bulletin* 2: 47–53.

Blank, Oliver, Bastian von Tresckow, Ina Monsef, Lena Specht, Andreas Engert, and Nicole Skoetz. 2017. Chemotherapy alone versus chemotherapy plus radiotherapy for adults with early stage Hodgkin lymphoma. *Cochrane Database of Systematic Reviews* 2017 (4): CD007110. https://doi.org/10.1002/14651858.CD007110.pub3.

Caldwell, Glyn G., Delle B. Kelley, and Clark W. Heath Jr. 1980. Leukemia among participants in military maneuvers at a nuclear bomb test. A preliminary report. *Journal of the American Medical Association* 244: 1575–1578.

Cannegieter, S.C., F.R. Rosendaal, and E. Briët. 1994. Thromboembolic and bleeding complications in patients with mechanical heart valve prostheses. *Circulation* 89: 635–641.

Cartwright, Nancy. 2007. Are RCTs the gold standard ? *BioSocieties* 2: 11–20.

Casey, Jonathan D., Daniel H. Solomon, Thomas A. Gaziano, Amy Leigh Miller, and Joseph Loscalzo. 2013. A patient with migrating polyarthralgias. *New England Journal of Medicine* 369: 75–80.

Cochrane. 2019. http://www.cochrane.org. Accessed 11 Feb 2019.

Criqui, Michael H., Melissa Austin, and Elizabeth Barrett-Connor. 1979. The effect of non-response on risk ratios in a cardiovascular disease study. *Journal of Chronic Diseases* 32: 633–638.

Doll, Richard, and A. Bradford Hill. 1950. Smoking and carcinoma of the lung. Preliminary report. *British Medical Journal* 221: 739–748.

Doll, Richard, Richard Peto, Jillian Boreham, and Isabelle Sutherland. 2004. Mortality in relation to smoking: 50 years' observations on male British doctors. *British Medical Journal* 328: 1519–1528.

Eddy, David M., and Charles H. Clanton. 1982. The art of diagnosis. Solving the clinicopathological exercise. *New England Journal of Medicine* 306: 1263–1268.

Glymour, Clark. 1980. *Theory and evidence*. Princeton: Princeton University Press.

Good, I.J. 1960. Weight of evidence, corroboration, explanatory power, information and the utility of experiments. *Journal of the Royal Statistics Society, Series B* 22: 319–331.

———. 1985. Weight of evidence: A brief survey. *Bayesian Statistics* 2: 249–270.

Gordis, Leon. 1996. *Epidemiology*. Philadelphia: W.B. Saunders.

Green, Stephanie, Jacqueline Benedetti, and John Crowley. 1997. *Clinical trials in oncology*. New York: Chapman and Hall.

Greenland, Sander, and Keith O'Rourke. 2008. Meta-analysis. In *Modern epidemiology*, ed. Kenneth J. Rothman, Sander Greenland, and Timothy L. Lash, 3rd ed., 652–682. Philadelphia: Lippincott, Williams and Wilkins.

Greenland, Sander, Kenneth J. Rothman, and Timothy L. Lash. 2008. Measures of effect and measures of association. In *Modern epidemiology*, ed. Kenneth J. Rothman, Sander Greenland, and Timothy L. Lash, 3rd ed., 51–70. Philadelphia: Lippincott, Williams and Wilkins.

Gurevitch, Jessica, Julia Koricheva, Shinichi Nakagawa, and Gavin Stewart. 2018. Meta-analysis and the science of research synthesis. *Nature* 555: 175–182.

Guyatt, Gordon H., Andrew D. Oxman, Gunn E. Vist, Regina Kunz, Yngve Falck-Ytter, and Holger J. Schünemann. 2008. GRADE: What is "quality of evidence" and why is it important to clinicians? *British Medical Journal* 336: 995–998.

Hernán, Miguel A., Sonia Hernández-Díaz, and James M. Robins. 2004. A structural approach to selection bias. *Epidemiology* 15: 615–625.

Higgins, Julian P.T., and Sally Green, eds. 2011. *Cochrane handbook for systematic reviews of interventions*. Version 5.1.0 [updated March 2011]. The Cochrane Collaboration. https://training.cochrane.org/handbook. Accessed 1 Apr 2019.

Ioannidis, John P.A. 2016. The mass production of redundant, misleading, and conflicted systematic reviews and meta-analyses. *The Milbank Quarterly* 94: 485–514.

Kassirer, Jerome P., John B. Wong, and Richard I. Kopelman. 2010. *Learning clinical reasoning*. 2nd ed. Baltimore: Lippincott Williams and Wilkins.

Klemetti, Anneli, and Lauri Saxén. 1967. Prospective versus retrospective approach in the search for environmental causes of malformations. *American Journal of Public Health* 57: 2071–2075.

Lalazar, Gadi, Victoria Doviner, and Eldad Ben-Chetrit. 2014. Unfolding the diagnosis. *New England Journal of Medicine* 370: 1344–1348.

MacMahon, Brian, Stella Yen, Dimitrios Trichopoulos, Kenneth Warren, and George Nardi. 1981. Coffee and cancer of the pancreas. *New England Journal of Medicine* 304: 630–633.

Manson, JoAnn E., Judith Hsia, Karen C. Johnson, Jacques E. Rossouw, Annlouise R. Assaf, Norman L. Lasser, Maurizio Trevisan, et al. 2003. Estrogen plus progestin and the risk of coronary heart disease. *New England Journal of Medicine* 349: 523–534.

Mayo, Deborah G. 2005. Evidence as passing severe tests: Highly probable versus highly probed hypotheses. In *Scientific evidence. Philosophical theories and applications*, ed. Peter Achinstein, 95–127. Baltimore: Johns Hopkins University Press.

Mueller, Nancy E. 1996. Hodgkin's disease. In *Cancer epidemiology and prevention*, ed. David Schottenfeld and Joseph F. Fraumeni Jr., 2nd ed., 893–919. Oxford: Oxford University Press.

Poole, Charles, Sander Greenland, Crystal Luetters, Jennifer L. Kelsey, and Gabor Mezei. 2006. Socioeconomic status and childhood leukaemia: A review. *International Journal of Epidemiology* 35: 370–384.

Rothman, Kenneth J. 1977. Epidemiologic methods in clinical trials. *Cancer* 39: 1771–1775.

Rothman, Kenneth J., Sander Greenland, and Timothy L. Lash. 2008. Validity in epidemiologic studies. In *Modern epidemiology*, ed. Kenneth J. Rothman, Sander Greenland, and Timothy L. Lash, 3rd ed., 128–147. Philadelphia: Lippincott, Williams and Wilkins.

Sackett, David L. 1979. Bias in analytic research. *Journal of Chronic Diseases* 32: 51–63.

Slone, Dennis, Samuel Shapiro, Lynn Rosenberg, David W. Kaufman, Stuart C. Hartz, Allen C. Rossi, Paul D. Stolley, and Olli S. Miettinen. 1978. Relation of cigarette smoking to myocardial infarction in young women. *New England Journal of Medicine* 298: 1273–1276.

Stampfer, Meir J., Graham A. Colditz, Walter C. Willett, JoAnn E. Manson, Bernard Rosner, Frank E. Speizer, and Charles H. Hennekens. 1991. Postmenopausal estrogen therapy and cardiovascular disease. Ten-year follow-up from the Nurses' Health Study. *New England Journal of Medicine* 325: 756–762.

Stegenga, Jacob. 2011. Is meta-analysis the platinum standard of evidence? *Studies in History and Philosophy of Biological and Biomedical Sciences* 42: 497–507.

Thun, Michael J., and S. Jane Henley. 2006. Tobacco. In *Cancer epidemiology and prevention*, ed. David Schottenfeld and Joseph F. Fraumeni Jr., 3rd ed., 217–242. Oxford: Oxford University Press.

Wynder, Ernest L., and Evarts A. Graham. 1950. Tobacco smoking as a possible etiologic factor in bronchiogenic carcinoma. *Journal of the American Medical Association* 143: 329–336.

Chapter 6
The Weight of Evidence Account Defended

Abstract In this chapter I defend the weight of evidence account. I argue that the other theories of evidence fail to adequately explain how evidence is gathered and used to confirm hypotheses in clinical medical science, and provide examples to show why they are unsuccessful. I argue that the weight of evidence account improves on the other theories by showing that it explains the many processes employed in clinical medical science, for example, by providing justification for more studies when the weight of evidence is low. The account also permits use of a variety of statistical methods and is not confined to frequentist or Bayesian approaches. It explains the case studies, and among other virtues it explains efforts to rank evidence, and justifies the use of treatment guidelines.

6.1 General

In this chapter, I will defend the weight of evidence account by arguing that the other accounts of evidence are unsatisfactory for clinical medical science, and that the weight of evidence account makes progress in remediation of deficiencies in the other accounts and also satisfactorily explains the case studies. I will also argue that the weight of evidence account explains the various efforts to rank evidence in clinical medical science, and that the ranks can be justified because they are based on the weight of evidence rationale.

6.2 Current Theories of Evidence Are Unsatisfactory for Clinical Medical Science

I have considered five theories of confirmation, in addition to Inference to the Best Explanation to the extent that it is considered a theory of confirmation or theory choice. None of these adequately explain how evidence is gathered and used in clinical medical science.

© Springer Nature Switzerland AG 2020
J. A. Pinkston, *Evidence and Hypothesis in Clinical Medical Science*, Synthese Library 426, https://doi.org/10.1007/978-3-030-44270-5_6

Popper's hypothetico-deductive method fails on several counts. His method relies on *modus tollens*, which is deductive. By refuting a prediction deduced from the hypothesis, the hypothesis is thereby refuted. A hypothesis surviving a test is "corroborated." Thus, the finding of a non-black raven refutes the hypothesis that all ravens are black. Popper realized that in the empirical sciences, statistical testing is widespread, and that statistical methods rely on probability models. As we have seen, the analysis of most hypotheses in therapeutic and etiologic research has relied on classical frequentist techniques.

Popper viewed statistical testing as deductive (1992, 432). A null statistical hypothesis may state, for example, that two treatments, A and B, are in some sense equivalent. An experiment can then be set up to test that hypothesis, and if test results show that the assumption of the null hypothesis is highly improbable, then it is rejected. The surviving hypothesis is then corroborated. Most RCTs and etiologic hypotheses are studied in this way. For RCTs in particular, these study methods do involve strategies to minimize or eliminate alternate hypotheses (from biases or confounding) that could explain the test outcomes, and thus may qualify as "severe tests" as envisioned by Popper.

Here it will be useful to distinguish what can be called a *scientific* hypothesis from a *statistical* hypothesis. A statistical hypothesis is rejected if study results show that it is highly improbable under the model assumptions. This is the method used to reject null hypotheses. Let us assume that an RCT has shown treatment A to be better in some sense than treatment B by its having led to the rejection of the null hypothesis of no difference in the treatments. Here the hypothesis that treatment A is better than treatment B is most likely the scientific hypothesis, and it was the scientific hypothesis that was being tested. Here it must be stressed that RCTs, particularly large ones, are not ordinarily undertaken unless preliminary studies or observations have indicated that A indeed might be better than B. The RCT is done to more vigorously test this hypothesis. Whereas the assumption underlying the statistical hypothesis is that the treatments are not different, the RCT is undertaken in fact to test the hypothesis that A is better than B.

Scientific hypotheses are the important hypotheses in clinical medical science; statistical hypotheses result from the assumptions of the statistical models and null statistical hypotheses are *assumed* as part of the model. The statistical tests are in essence tools used to test and evaluate scientific hypotheses. Is A really better than B, or could the results just be due to chance?

Although if in fact A is better than B, and the results of several RCTs support that hypothesis, a subsequent RCT finding no difference cannot disconfirm the hypothesis, because an RCT finding no difference not only *could* occur but even *is expected to occur* with some frequency, depending on the statistical model that is used. These considerations apply as well to etiologic hypotheses. Consider the studies of coffee drinking and pancreatic cancer (Anderson et al. 2006): Many studies found no association, some studies found an association, and some studies were equivocal. Studies that show no association in etiologic or therapeutic research cannot necessarily be relied on to reject the scientific hypothesis, thus a theory of evidence focusing on rejection of hypotheses at best explains only part of the relation between evidence and hypothesis in clinical medical science.

Another reason that Popper's approach is unsatisfactory is that he denied that hypotheses have probabilities, and he endeavored to show that his notion of corrob-oration is also not a probability. But in clinical medical science, some hypotheses clearly *do* have probabilities. This is true for many screening tests, and many diagnostic hypotheses as well. Consider the screening test for TB using chest x-rays in Table 3 (Yerushalmy et al. 1950; Brown and Hollander 1977). Here Bayes' theorem was used to show that someone with a positive chest x-ray reading had a higher probability of having TB than someone drawn randomly from the population, hence these hypotheses clearly have probabilities. And recall the way Bayes' theorem was used in Table 4.8 (Kassirer et al. 2010), where the various probabilities associated with the diagnoses under consideration were used to deter-mine the hypothesis with the highest probability, based on various findings. It is unclear whether Popper would object to these uses of probabilities assigned to hypotheses when they can be given a relative frequency interpretation, but Bayesian statistical methods are also occasionally used in clinical medical science, and these are not given a frequentist interpretation (discussed more below). Thus Popper's methods are untenable here as well.

Hempel's "satisfaction theory" is based on logical relations between hypotheses and confirmatory observations, thus is ill – suited to the probabilistic manner in which hypotheses are confirmed in clinical medical science (Hempel 1945a, b). It allows for the derivation of absurdities and "grue" hypotheses (Goodman 1983), as previously noted, since it is based only on the syntax of the sentences expressing the evidence and hypothesis, considering the relation to be two – place and ignoring background information (Earman 1992, 68–75; Howson and Urbach 2006, 299).

Bayesian confirmation theory also fails because, it will be recalled, the theory requires that hypotheses have probabilities. Evidence *e* confirms hypothesis *h* just to the extent that *e* raises the probability that *h* enjoyed before the acquisition of *e* (its prior probability) to *h*'s probability after the acquisition of *e* (its posterior probabil-ity). Since frequentist statistics are used for analysis in the great majority of thera-peutic and etiologic studies, the results of these studies do not have a Bayesian interpretation. Neither the scientific nor the statistical hypotheses have probabilities. The rejection of the null hypothesis does not result in the null hypothesis receiving a low probability, nor does the scientific hypothesis (the "alternate" hypothesis under the model) receive a higher probability.

As we have seen however, Bayes' theorem is used in some areas of clinical medical science, including screening tests and in medical diagnoses. But the use of Bayes' theorem *per se* is not an endorsement of Bayesian epistemology, since Bayes' theorem is derivable from the axioms of probability theory and is used by frequentist as well as Bayesian statisticians. And, as Glymour (2010, 334) notes, Bayesian statistics and Bayesian epistemology are different. In clinical medical science, the use of Bayes' theorem ordinarily requires the use of a "gold standard," or criteria against which persons having some characteristic such as a positive chest x-ray finding or laboratory result can be assigned a probability of having the diagnosis in question. In the example of using chest x-rays as a screening test for TB in Table 3, for example, the definitive diagnosis of TB would require the gold standard of a positive direct sputum examination or sputum culture revealing the presence of Tubercle bacilli. The use of Bayes' theorem in Bayesian epistemology is not given a frequentist interpretation.

Although Bayesian statistical analysis of the types of studies found in clinical medical science is possible, including RCTs (Cox et al. 2014, 2351; Berry 2006), by far most are analyzed using frequentist methods. Some defenders of the Bayesian theory of confirmation have derided the use of frequentist statistical analysis. For example, Howson and Urbach state, "Classical estimation theory and significance tests . . . are not estimates in any normal or scientific sense, and, like judgments of "significance" and "non-significance", they carry no inductive meaning at all. Therefore, they cannot be used to arbitrate between rival theories or to determine practical policy" (2006, 181–182). But, in clinical medical science, classical, frequentist techniques *are* used to arbitrate between rival theories and to determine practical policy, such as their use in establishing treatment guidelines.

The most important aspect of a scientific study, I am arguing, is its *accuracy*. The science of statistics is itself evolving, and it seems premature to insist that some particular statistical approach should be adopted to the exclusion of others. I will have more to say on these matters below and in Chap. 7 where I will defend randomization and frequentist statistical analysis against Bayesian and other critics.

Mayo's error-statistical theory of confirmation (1996) is also unsatisfactory for clinical medical science, notwithstanding the fact that classical frequentist statistical methods are almost exclusively used in the analysis of therapeutic and etiologic hypotheses, methods that she champions. Her error-statistical theory of experiment does a satisfactory job of describing therapeutic research, which is mostly experimental and in which studies are usually designed to have sufficient numbers of subjects to detect differences in outcome at or below some threshold significance level (e.g., $\alpha < .05$) and sufficient power to justify the inference that a false null statistical hypothesis has been rejected. Many such studies are RCTs in which randomization and stratification are used to minimize errors, thus constituting severe tests according to her theory; thus, a given hypothesis would pass such a test if it met her probability requirement. Her theory of experiment applies well to individual RCTs, or even to non-randomized clinical trials that have an appropriate control group.

However, her theory does not illuminate other important areas in therapeutic research, such as how many RCTs should be done in some particular case except to occasionally imply that sometimes more tests may be needed. Her philosophy of experiment may extend to meta-analysis, which is a method of analysis that is also based on frequentist statistical methodology, but is silent on such things as systematic reviews of RCTs and other studies not employing meta-analyses, which are pivotal to the notion of weight of evidence that I am developing.

However, even if her theory is correct, much more is needed, I am arguing, for a satisfactory theory of evidence for clinical medical science. Issues concerning error and accuracy in etiologic research, which is mostly observational, and in the methods that underpin the confirmation of diagnostic hypotheses must be addressed. For etiologic hypotheses, for example, a description and explanation of the types of biases that may occur, as well as confounding, is needed. A discussion and explanation of how diagnostic hypotheses are confirmed, with attention to the sources of inaccuracy that may arise, is also required.

Achinstein's theory of evidence (2001) is also unsatisfactory. As we have seen, it includes both subjective and objective notions, the latter he claims being of most importance to scientists. His objective notion of potential evidence, it will be recalled, includes that

p(there is an explanatory connection between h and $e/e\&b$) > .5

His theory requires that h be assigned a probability, and is similar to Bayesianism in this respect. Since the great majority of therapeutic and etiologic hypotheses are analyzed using frequentist statistical methods, this renders study results not interpretable in his schema. Thus, for example, if I hold 800 tickets in a fair 1000 ticket lottery, it is consonant with Achinstein's notion of objective epistemic probability that it is four times more reasonable to believe that I will win than that I will not win, and it qualifies as his potential evidence that I will win. But if I examine the results of one or more studies of the type found in therapeutic and etiologic research, I seem to be in a quandary if I adopt this notion. For example, suppose that after examining the results of the British doctor study on cigarette smoking and lung cancer (Doll et al. 2004), I am impressed by the substantially increased frequency of lung cancer cases among smokers compared with non-smokers that cannot reasonably be explained by chance. How many more times is it reasonable to believe that cigarette smoking causes lung cancer than that it does not? If this is potential evidence, what is the resulting probability?

Another shortcoming of Achinstein's approach is that for e to be evidence that h, e must supply a "good reason to believe" h, and that one of his necessary requirements is that p(h/e) > .5, since e cannot also supply a good reason to believe ~h. Recall that one of his objections to Bayesianism was that for e to be evidence that h, all that e must do is to increase h's probability. But if the Bayesian concept of evidence is too weak, Achinstein's is arguably too strong for a desired concept of evidence in clinical medical science. In differential diagnosis, for example, the presence of fever surely would be considered evidence of an infection, although in some particular case if that were the only evidence then its weight for that hypothesis might be relatively weak. But it still would be *evidence*. Consider the case of meeting diagnostic criteria for a specific diagnosis, as in the example of infective endocarditis (Casey et al. 2013). Recall that to make the diagnosis of infective endocarditis, it was necessary to meet the Duke criteria. The diagnosis is established when two major, one major plus three minor, or five minor criteria are met. Each individual major and minor criterion is evidence for the diagnosis, but no single criterion alone establishes it. But any one of the criteria should qualify as evidence. Also, in the TB and chest x-ray example of Table 3, following a positive x-ray reading the probability of having TB was raised from its prior probability of .016 to the posterior probability of .30. This should be considered evidence of having TB, although it does not raise the probability of having TB to more than .5. Thus, for the above reasons, Achinstein's theory falls short here as a satisfactory theory of evidence for clinical medical science.

Many diagnostic hypotheses and their supporting evidence, however, are probably interpretable under Achinstein's theory. For example, as Eddy and Clanton (1982) have noted, when several diagnostic hypotheses are considered, some are eliminated as being less probable and the most probable becomes the working diagnosis. This kind of "eliminative induction" would usually result in an initial probability of somewhere near .5 or greater, and often the other conditions for potential evidence would be met. But since the weight of evidence account that I am developing need not be a probability, it has no clear or necessary correspondence to Achinstein's notion of objective epistemic probability.

It might also be worth noting that his notion of subjective evidence may also often be applicable. Subjective evidence, it will be recalled, satisfies at least three conditions, where X is a person or group (Achinstein 2001, 23):

1. X believes that e is evidence that h
2. X believes that h is true or probable
3. X's reason for believing that h is true or probable is that e is true

Here, e does not need to be true, but only that X believes it is. Evidence e is accepted until new evidence refutes it. It requires belief, and that someone or some group is in a certain epistemic situation in regard to the evidence. Thus, this notion of subjective evidence seems to fit the situation that exists with most hypotheses in clinical medical science that are supported by a substantial weight of evidence, for example, the hypothesis that cigarette smoking causes lung cancer.

Before leaving this section, it is worthwhile to further emphasize the centrality of *accuracy* to the weight of evidence account, and to stress that this notion of accuracy enables this account to apply to hypotheses concerning both individuals and larger, empirical hypotheses involving groups. Accuracy combines validity and precision, and validity bears on the question of whether we are measuring what we intend to measure. For example, for the medical diagnostician, are we actually measuring antibodies to a specific virus in this blood test? For the RCT investigator, are we actually measuring the effects of treatments A and B in these groups? For the epidemiologist, are we actually measuring the effect of exposure x on the frequency of disease y in this population? To the extent that we can be confident that the answer is yes, then confidence in validity is increased. The degree of validity combined with the degree of precision (precision is often associated with confidence intervals in statistical analysis) yields the degree of accuracy.

Thus, evidence associated with an individual, for example the N of 1 study, a diagnostic hypothesis, or a blood test, is equally interpretable in the weight of evidence account with evidence derived from larger, empirical hypotheses that are studied in groups. What links them is that the same notion of accuracy is applied to both. Thus, these considerations undergird my arguments, including that my notion of accuracy is a better measure than Mayo's severity, since Mayo's theory deals with experimental studies but validity (hence accuracy) is of more pressing concern in observational studies where experimental control is lacking.

6.3   The Weight of Evidence Account Remedies Deficiencies in Other Accounts

Compared with previous accounts, the weight of evidence account better explains the way that evidence is gathered and used to confirm the types of hypotheses encountered in clinical medical science, and represents progress in the remediation of deficiencies in those accounts. For example, I am arguing that the notion of *accuracy* is more salient here than is the notion of "severity" advanced by Popper and Mayo. Since Popper's notion is best understood as being based on comparative likelihoods, it is not directly applicable to the type of frequentist statistical testing typically done in clinical medical science. Mayo has argued that passing a severe test is more probative in securing evidence for a hypothesis than is increasing the probability of the hypothesis, which she has referred to as "the highly probed versus highly probable debate" (2005, 96). Her notion of severity is quantified as 1-α, where the α level is the result of a frequentist statistical test. I am advancing the notion of *accuracy* as being the more important concept, not just for assessing an individual observation or study, but also groups of these. The weight of evidence account with its focus on accuracy improves on the severity notion and at the same time embraces probability as a measure of weight of evidence for those hypotheses that have probabilities. In a well designed and executed experiment like an RCT or other controlled study, a result with a low α level, say $\alpha < .05$, constitutes passing a severe test under Mayo's theory, and also would be considered an accurate result and accorded strong evidence under the weight of evidence account.

The weight of evidence account explains the need for more than one study, especially in cases where the threat to accuracy is high. Severity applied to individual tests or even to a meta-analysis does not explain this. For a given hypothesis, how many studies should be done? The other accounts do not adequately address this, but the answer I would argue is, in general, that the number is sufficient when additional studies are not thought to appreciably change the weight of evidence. Thus it is unnecessary, for example, to do another etiological study of cigarette smoking and lung cancer since the present weight of evidence is sufficiently strong to establish a causal relationship. Likewise, unless methodological advances are made to better rule out confounding, further studies are probably unnecessary on coffee drinking as a possible cause of pancreatic cancer, since a study showing a statistically significant result will likely be interpreted as the result of residual confounding in view of the large number of negative studies already done.

The weight of evidence account accommodates hypotheses with probabilities, where the weight of evidence is directly related to the probability. It explains why a working diagnosis is accorded a lesser weight of evidence, since it is less probable than a diagnosis meeting a gold standard or diagnostic criteria, which would establish a high probability and maximal weight of evidence. How many observations are needed in a particular case of differential diagnosis? This number will vary, depending on a number of factors, one of the most important being the severity of the disease and the need for a correct diagnosis. For example, for the common cold, the presence of fever, sore throat, and nasal congestion may be sufficient for diagnosis and treatment; for someone brought to the emergency department in a coma, many more observations will be needed, including more detailed physical examination, radiographic and laboratory testing, and the like.

The weight of evidence account is superior also because it allows for a multiplicity of statistical approaches and does not depend on a particular interpretation of statistics or statistical theory. Thus frequentist, Bayesian, likelihoodist, or other methods may be used depending on the type of study and method of analysis considered most appropriate by the investigators. Frequentist methods are at present used most commonly, but Bayesian methods are also sometimes employed and may even be preferable in some cases, such as bias analysis. As I have argued, the science of statistics is itself evolving, and new methods are being developed and tested. Modern clinical medical science uses statistics as an inferential tool in the assessment of scientific hypotheses, and more than one method may be applicable in any given case. Indeed, Mayo (1996, 69) has acknowledged that given a set of well-defined statistical problems, and for given sets of data, Bayesian and non-Bayesian inferences may formally agree, notwithstanding differences in rationale and interpretation. This inclusiveness of the weight of evidence account overcomes the limitations of Mayo's frequentist-based approach, and also overcomes the limitations imposed by Bayesianism and Achinstein's theory due to their requirement that hypotheses have probabilities.

6.4   The Weight of Evidence Account Explains the Case Studies

The weight of evidence account explains why the evidence in the case studies is accorded the degree of confidence by clinical medical scientists that it has received. I am arguing that scientists have confidence in the results of observations or studies just to the extent that they are judged to be accurate, and this applies to a single observation or study or to groups of these.

For therapeutic hypotheses, recall the RCT performed by the Gastrointestinal Tumor Study Group (Thomas and Lindblad 1988) that randomized patients with rectal cancer to surgery only, surgery plus radiotherapy, surgery plus chemotherapy, or to surgery plus both chemotherapy and radiotherapy. The results showed a statistically significant benefit for the group that received both radiotherapy and chemotherapy after surgery. The RCT maximized accuracy due to the study design and randomization procedure that in principle minimized or eliminated any effect of bias or confounding. Thus the study provided relatively accurate evidence for the result that was quantified by the statistically significant p-values associated with the outcome measures of effect. And to reiterate, the authors emphasized that the control group (surgery only) had a much better outcome than would have been expected based on historical groups of similar patients treated only with surgery, illustrating the need for concurrent controls in randomized studies, and thus why RCTs provide the most accurate source of evidence.

The weight of evidence account also explains why evidence from an N of 1 trial receives a greater weight of evidence for the treatment of an individual patient than evidence from a study of a group of patients with the same condition. Consider, for example, a hypothetical study of pain reduction by a drug in patients experiencing migraine headache. Suppose a study showed an average of 60% pain reduction (95% C.I. 40–80%) in a group of 200 patients receiving the drug, but an N of 1 trial in a patient showed a 30% average reduction with a range of 25–35%. The predictive value from the N of 1 trial is more accurate for that individual than is the average and range derived from the study of the group, since the results are averaged over a large number of individuals in the group. This is what occurred in the N of 1 trial of theophylline in the patient with asthma in Chap. 4 (Guyatt et al. 1986), in which the patient felt worse when taking the drug. Theophylline would have been predicted to be beneficial based on studies in groups of patients with asthma.

The weight of evidence account also satisfactorily explains the degree of evidential support accorded to the etiologic hypotheses illustrated by the case studies. Because of the strong effect of cigarette smoking on the diseases studied, the weight of evidence was strong for a causal role for the subsequent development of chronic obstructive lung disease, ischemic heart disease, and cancers of the lung, esophagus, and upper aerodigestive tract as indicated by the results of the large British doctor study (Doll et al. 2004) and supported by the case-control studies. In addition, a causal role for cigarette smoking and the subsequent occurrence of myocardial infarction in healthy young women was found in the case-control study by Slone et al. (1978), and a causal role for cigarette smoking in pancreatic cancer was reported by the 2004 IARC review, which concluded that the evidence was strong. A strong association was found between maternal stilbestrol use and later vaginal adenocarcinoma in young female offspring (Herbst et al. 1971).

The main reason that the weight of evidence is strong for the above etiologic hypotheses is because of the strength of the association between the exposure and the disease. Even though etiologic studies are more prone to inaccuracy, the strength of the associations allowed any effect from bias or confounding to be overcome. However, as we have seen, when associations between exposure and disease are weak or non-existent, etiologic study methodologies often give conflicting results. Thus the relative risks for an association between tonsillectomy and later Hodgkin lymphoma ranged from 0.7 to 3.6 (Mueller 1996), and it was concluded that it was unlikely that tonsillectomy was a risk factor for Hodgkin lymphoma in young or middle aged adults, but it was still unclear whether it was a risk factor for the disease later in life (Mueller et al. 1987). Similarly, the great variation in the results of studies of coffee drinking and cancer of the pancreas (Anderson et al. 2006) led to the conclusion that an association is unlikely, and that studies showing an association were probably confounded.

In the case of the cross-sectional study of a possible relation between alcohol consumption and systolic blood pressure (Ueshima et al. 1984; Kelsey et al. 1986), it will be recalled that the variables of age, alcohol consumption, uric acid, and ponderosity index among the Osaka men accounted for only about 18% of the total variation in systolic blood pressure, indicating that probably important risk factors were left out of the model. As this case illustrates, one problem with the analysis of cross-sectional studies when used for etiologic research is that the investigators must gather information on the variables thought to be of etiologic interest, and important risk factors for outcome may be left out. Coupled with the problems associated with the temporal relation of exposure to disease and with measuring prevalence rather than incidence identified earlier, this relative lack of accuracy renders results from cross-sectional studies as a group to provide a lower weight of evidence when etiologic hypotheses are considered than do the cohort and case-control study designs.

The weight of evidence account explains why diagnostic hypotheses are accorded weights of evidence based on probabilities of being correct. The probabilities involved can be more or less well quantified depending on the information available. For example, in the chest x-ray and TB screening case illustrated in Table 3, a positive x-ray reading raised the probability of having TB from .016 to .30 (Yerushalmy et al. 1950; Brown and Hollander 1977). Thus, the evidence arguably went from very weak to perhaps weak-to-moderate. Contrast this with the evidence presented in Table 4.8 of the renal insufficiency case (Kassirer et al. 2010) in which the probability of atheromatous embolism went from .01, which might be characterized as negligible or very weak, to .977, which must be considered very strong.

The weight of evidence account explains why working diagnoses are accorded a lower weight than diagnoses that have met a gold standard or diagnostic criteria. In the case of the young woman with fever, elevated white cell count, and arthralgias presented in Chap. 4 (Casey et al. 2013), the initial observations were non-specific and could not differentiate between infectious and noninfectious inflammatory causes. The working diagnosis was "viral infection," and she was prescribed anti-inflammatory drugs. At this point, the weight of evidence for infection was at best somewhere around .5. On her next visit to the emergency department, she was found to have a heart murmur. The new heart murmur and recent fever narrowed the differential diagnosis, and infective endocarditis became a new diagnostic consideration. The heart murmur and positive blood cultures would have established infective endocarditis as the diagnosis since they are both major Duke criteria, and would have met criteria for the diagnosis. The blood cultures were negative, however, and it will be recalled that she subsequently met the Jones criteria for the diagnosis of acute rheumatic fever, which was the final diagnosis.

In the above case the initial working diagnosis of viral infection gave way to the diagnosis of infective endocarditis or acute rheumatic fever as the disease progressed and new observations were made. The negative blood cultures provided evidence against infective endocarditis and sheds light on the notion of accuracy in differential diagnosis. Blood cultures are usually obtained by taking several independent blood samples (typically at least three) under sterile conditions in order to reduce the chance of bacterial contamination. This illustrates an example of a threat to accuracy, in this case validity, since a false positive blood culture could lead to misdiagnosis. The practice of multiple separate samples reduces this threat. An example of a precision issue is that of the patient's recollection of shortness of breath as a child, consistent with rheumatic fever in childhood. But "shortness of breath" *per se* is imprecise. Her description of getting short of breath easily and not being able to play with the other children provided enough evidence that her shortness of breath was of sufficient magnitude to be a factor in the diagnosis of rheumatic fever. And although this patient met two major Jones criteria (migratory arthritis and carditis) and one minor criterion (fever) which established the diagnosis of rheumatic fever, additional evidence was sought which included blood tests for antibodies showing previous streptococcal infection, which were positive, thus strengthening confidence that the diagnosis was the correct one.

The rheumatic fever case illustrates the process that is most frequently used to confirm a diagnostic hypothesis. No probabilities were calculated. Accuracy here is focused on the accuracy of diagnostic maneuvers including clinical observations, testing results, and the like. No effort is made to accurately estimate probabilities on the way to making a diagnosis. The analysis by Eddy and Clanton (1982) seems relevant here. The finding of a heart murmur might accord well with their notion of a *pivot*: a finding around which the various diagnostic hypotheses coalesce. What explains fever, arthralgias, and a new heart murmur? Evidence is then sought to discount some causes of these findings and increase the probability of others, but it is done as a series of comparisons and no probabilities are calculated. A better description is that the weight of evidence for or against these hypotheses shifts as evidence for or against them accumulates. This is particularly the case when a disease may be evolving in its manifestations. Any argument that the process is one that necessarily involves an explicit consideration of probabilities would seem to leave the proponent as expressing some notion of subjective probability. While this might fit with at least some versions of subjective Bayesianism, it would seem that this would represent a less satisfactory approach than the weight of evidence account, which I argue subsumes this line of reasoning (Bayesianism in general) into a more comprehensive framework.

6.5   The Weight of Evidence Account Explains Efforts to Rank Evidence

The weight of evidence account explains the various efforts to rank evidence in clinical medical science. For example, consider the *GRADE* system for rating the quality of evidence (Guyatt et al. 2008). In this system, the quality of evidence reflects the extent to which confidence in an estimate of effect is adequate to support recommendations such as those that are found in treatment guidelines. The approach considers study design to be important in grading the quality of evidence, and avers that RCTs in general provide stronger evidence than do observational studies. Factors that can decrease the quality of evidence include study limitations (e.g., assessments being biased), inconsistency of results across studies, and imprecision. Factors that can increase the quality of evidence include a large magnitude of effect and a dose-response gradient. The evidence supporting a recommendation is graded according to quality, and the authors provide several examples in which quality is assessed as being of high quality to very low quality.

This system is satisfactorily explained as being focused on accuracy in the evidence that is assessed, and the weight of evidence account satisfactorily explains their grading of evidence according to what they consider quality.

Another example is that provided by the NCCN (2015) in the assessment of evidence in developing guidelines. The NCCN is a not-for-profit alliance of 26 mostly university-affiliated cancer centers located throughout the U.S. Expert panels are assembled and constituted similarly to those of Cochrane and the IARC. Among NCCN publications are detailed guidelines for the treatment of cancer according to type or tissue of origin (e.g., breast, lung) and factors such as stage of disease. A separate expert panel exists for each area (e.g., Hodgkin lymphoma).

The NCCN categories of evidence are "high-level" and "lower-level." The level of evidence depends on the extent of data (e.g., number and size of trials), consistency of data (similar or conflicting results), and quality of data (e.g., RCTs, non-RCTs, meta-analysis or systematic reviews, clinical case reports, case series). It seems clear that in assessing these factors that the notion of accuracy as it applies to each observation (e.g., a clinical case report) or study, or groups of these, satisfactorily explains the assignment of evidence into the high-level and lower-level categories. Also, the categories themselves are readily explained as efforts to assign different weights to the evidence.

The weight of evidence account also explains the "hierarchical pyramid" of EBM, which is taken up in the next chapter.

References

Achinstein, Peter. 2001. *The book of evidence*. Oxford: Oxford University Press.
Anderson, Kristin E., Thomas M. Mack, and Debra T. Silverman. 2006. Cancer of the pancreas. In *Cancer epidemiology and prevention*, ed. David Schottenfeld and Joseph F. Fraumeni Jr., 3rd ed., 721–762. Oxford: Oxford University Press.
Berry, Donald A. 2006. Bayesian clinical trials. *Nature Reviews Drug Discovery* 5: 27–36.
Brown, Byron Wm., Jr., and Myles Hollander. 1977. *Statistics: A biomedical introduction*. New York: Wiley.
Casey, Jonathan D., Daniel H. Solomon, Thomas A. Gaziano, Amy Leigh Miller, and Joseph Loscalzo. 2013. A patient with migrating polyarthralgias. *New England Journal of Medicine* 369: 75–80.
Cox, Edward, Luciana Borio, and Robert Temple. 2014. Evaluating ebola therapies – The case for RCTs. *New England Journal of Medicine* 371: 2350–2351.

Doll, Richard, Richard Peto, Jillian Boreham, and Isabelle Sutherland. 2004. Mortality in relation to smoking: 50 years' observations on male British doctors. *British Medical Journal* 328: 1519–1528.

Earman, John. 1992. *Bayes or bust? A critical examination of Bayesian confirmation theory.* Cambridge: MIT Press.

Eddy, David M., and Charles H. Clanton. 1982. The art of diagnosis. Solving the clinicopathological exercise. *New England Journal of Medicine* 306: 1263–1268.

Glymour, Clark. 2010. Explanation and truth. In *Error and inference: Recent exchanges on experimental reasoning, reliability, and the objectivity and rationality of science*, ed. Deborah G. Mayo and Aris Spanos, 331–350. Cambridge: Cambridge University Press.

Goodman, Nelson. 1983. *Fact, fiction, and forecast.* 4th ed. Cambridge: Harvard University Press.

Guyatt, Gordon, David Sackett, D. Wayne Taylor, John Chong, Robin Roberts, and Stewart Pugsley. 1986. Determining optimal therapy – Randomized trials in individual patients. *New England Journal of Medicine* 314: 889–892.

Guyatt, Gordon H., Andrew D. Oxman, Gunn E. Vist, Regina Kunz, Yngve Falck-Ytter, and Holger J. Schünemann. 2008. GRADE: What is "quality of evidence" and why is it important to clinicians? *British Medical Journal* 336: 995–998.

Hempel, Carl G. 1945a. Studies in the logic of confirmation (I). *Mind* 54: 1–26. Reprinted, with some changes, in Hempel 1965, 3–51.

———. 1945b. Studies in the logic of confirmation (II). *Mind* 54: 97–121. Reprinted, with some changes, in Hempel 1965, 3–51.

———. 1965. *Aspects of scientific explanation and other essays in the philosophy of science.* New York: The Free Press.

Herbst, Arthur L., Howard Ulfelder, and David Poskanzer. 1971. Adenocarcinoma of the vagina. Association of maternal stilbestrol therapy with tumor appearance in young women. *New England Journal of Medicine* 284: 878–881.

Howson, Colin, and Peter Urbach. 2006. *Scientific reasoning. The Bayesian approach.* 3rd ed. Chicago: Open Court.

International Agency for Research on Cancer. 2004. *Tobacco smoke and involuntary smoking. IARC monographs on the evaluation of carcinogenic risks to humans.* Vol. 83. IARC: Lyon.

Kassirer, Jerome P., John B. Wong, and Richard I. Kopelman. 2010. *Learning clinical reasoning.* 2nd ed. Baltimore: Lippincott Williams and Wilkins.

Kelsey, Jennifer L., W. Douglas Thompson, and Alfred S. Evans. 1986. *Methods in observational epidemiology.* Volume 10 of *Monographs in epidemiology and biostatisics*, general ed. Abraham M. Lilienfeld. Oxford: Oxford University Press.

Mayo, Deborah G. 1996. *Error and the growth of experimental knowledge.* Chicago: University of Chicago Press.

———. 2005. Evidence as passing severe tests: Highly probable versus highly probed hypotheses. In *Scientific evidence. Philosophical theories and applications*, ed. Peter Achinstein, 95–127. Baltimore: Johns Hopkins University Press.

Mueller, Nancy E. 1996. Hodgkin's disease. In *Cancer epidemiology and prevention*, ed. David Schottenfeld and Joseph F. Fraumeni Jr., 2nd ed., 893–919. Oxford: Oxford University Press.

Mueller, Nancy, G. Marie Swanson, Chung-cheng Hsieh, and Philip Cole. 1987. Tonsillectomy and Hodgkin's disease: Results from companion population-based studies. *Journal of the National Cancer Institute* 78: 1–5.

NCCN (National Comprehensive Cancer Network). 2015. http://www.nccn.org. Accessed 5 Dec 2015.

Popper, Karl. 1992. *The logic of scientific discovery.* New York: Routledge.

Slone, Dennis, Samuel Shapiro, Lynn Rosenberg, David W. Kaufman, Stuart C. Hartz, Allen C. Rossi, Paul D. Stolley, and Olli S. Miettinen. 1978. Relation of cigarette smoking to myocardial infarction in young women. *New England Journal of Medicine* 298: 1273–1276.

Thomas, Patrick R.M., and Anne S. Lindblad. 1988. Adjuvant postoperative radiotherapy and chemotherapy in rectal carcinoma: A review of the gastrointestinal tumor study group experience. *Radiotherapy and Oncology* 13: 245–252.

Ueshima, Hirotsugu, Takashi Shimamoto, Minoru Iida, Masamitsu Konishi, Masato Tanigaki, Mitsunori Doi, Katsuhiko Tsujioka, et al. 1984. Alcohol intake and hypertension among urban and rural Japanese populations. *Journal of Chronic Diseases* 37: 585–592.

Yerushalmy, J., J.T. Harkness, J.H. Cope, and B.R. Kennedy. 1950. The role of dual reading in mass radiography. *The American Review of Tuberculosis* 61: 443–464.

Chapter 7
Justification for the Hierarchical Pyramid of Evidence-Based Medicine and a Defense of Randomization

Abstract In this chapter I argue that the weight of evidence account justifies the "hierarchical pyramid" of study types often used by the Evidence-Based Medicine movement to rank evidence according to the degree of evidential support they afford. I also defend the need for randomization in randomized clinical trials against critics from the medical and philosophical communities, and explain why accuracy is thereby improved. I illustrate the process of how studies that are early and inconclusive progress to more definitive studies by considering the historical case of the evolution of treatments for early breast cancer. I argue that the weight of evidence account successfully explains why the earlier, lower level studies were insufficient and later, more definitive studies were necessary. I also argue that studies in clinical medical science are frequently more generalizable than critics seem to often assume.

7.1 General

No doubt one of the most important aims of the EBM movement is to place the clinical practice of medicine on a firm scientific evidential basis. According to Sackett et al., "Evidence-based medicine is the conscientious, explicit, and judicious use of current best evidence in making decisions about the care of individual patients" (1996, 71). EBM requires the integration of the best research evidence with the clinician's clinical expertise and judgment and each patient's unique values and circumstances (Straus et al. 2011, 1). Greenhalgh has emphasized the quantitative nature of most modern clinical medical research, and proposes an alternative definition of EBM:

> Evidence-based medicine is the use of mathematical estimates of the risk of benefit and harm, derived from high-quality research on population samples, to inform clinical decision-making in the diagnosis, investigation or management of individual patients (2010, 1).

The EBM community postulates that evidence derived from some types of studies provide, in general, better confirmation of the hypotheses under test than do other types of studies. This is often illustrated in the form of a "hierarchical pyramid," with systematic reviews of randomized clinical trials and meta-analyses of these trials at the

© Springer Nature Switzerland AG 2020

J. A. Pinkston, *Evidence and Hypothesis in Clinical Medical Science*, Synthese Library 426, https://doi.org/10.1007/978-3-030-44270-5_7

top, or apex of the pyramid, and case studies, anecdote, bench studies, personal opinion and the like at the bottom, or base (Greenhalgh 2010, 18). One suggested ordering, from highest to lowest, of the relative weight carried by the different types of primary study when making decisions about clinical interventions is (Greenhalgh 2010, 43–44)[1]:

1. Systematic reviews and meta-analyses
2. RCTs with definitive results (i.e. confidence intervals which do not overlap the threshold clinically significant effect)
3. RCTs with non-definitive results (i.e. a point estimate which suggests a clinically significant effect but with confidence intervals overlapping the threshold for this effect
4. Cohort studies
5. Case-control studies
6. Cross-sectional surveys
7. Case reports

The above ordering of "relative weights" to be given to types of studies suggest that what might be termed "levels" of evidence may exist that bear on the confirmation of hypotheses. For example, "suggestive" evidence from some type of study that some hypothesis may be superior to alternatives (lower-level evidence) may encourage more extensive and focused research to gather more confirmatory evidence (higher-level evidence). This notion of lower-level evidence prompting acquisition of higher-level evidence in the pursuit of stronger confirmation can be illustrated by studies that were performed in the evolution of treatments for early breast cancer.

7.2   The Evolution of Treatments for Early Breast Cancer

7.2.1   Background

Cancer of the female breast begins as an uncontrolled proliferation of cells lining the breast milk ducts or lobules and spreads locally, regionally, and distantly. Local spread occurs in the breast itself where the tumor grows into a mass or lump that can become quite large and even ulcerate the overlying skin. Regional spread refers to cancer cells entering the lymph channels that drain to lymph glands located in the axilla, base of the neck just above the clavicle, and sometimes to glands located along the sternum. Distant spread occurs mainly by cancer cells moving through lymph channels or invading small blood vessels and being carried to other organs where they may

---

[1]Greenhalgh cites Guyatt et al. (1995) in connection with this listing. Also, as she notes, this schema does not include qualitative research, which also may be of importance (Greenhalgh 2010, 163–176).

lodge and grow. The most common sites of such spread are the bones, lungs, liver, and brain. Early breast cancer, in which the cancer is still confined to the breast, or even in some cases of regional spread, is potentially curable. Breast cancer that has spread to distant organs is almost invariably incurable.

Modern treatment for women with early breast cancer consists mainly of surgery, radiation therapy, and systemic therapies including drugs, antibodies, or hormonal therapy. These modalities may be used alone or in some combination. The curability of early breast cancer is achieved primarily by surgery and/or radiation therapy, and much interest has been shown in trying to determine how best to use these forms of treatment in order to achieve the highest probability of cure, and at the same time minimize the adverse effects of the treatment itself.

Physicians have used surgery in some form or other for centuries. In 1867, Charles Moore, an English surgeon, introduced an extensive operation for the removal of cancer of the breast (Williams et al. 1953). Experimentation and refinement of technique led to an operation called the *radical mastectomy*, usually associated with the English surgeon W.S. Halsted. In the radical mastectomy, the entire breast is removed, along with the underlying muscles and muscles around the axilla. Most of the axillary contents on the side where the cancer is located are removed. The operation is psychologically damaging and physically disfiguring, and swelling of the arm occurs frequently due to the disruption of lymphatic channels from the axillary dissection. The objective is to radically remove all cancer. More aggressive variations of the Halsted radical mastectomy involve removal of the lymph glands around the clavicle and along the sternum.

Röntgen discovered X-rays in 1895, and experimentation with the new rays for medical applications was swift. By the 1920s, high voltage irradiation for the treatment of cancer was introduced (Adair 1943). Most early experimentation with X-ray treatment was by trial and error as physicians learned techniques for maximizing the antitumor effects while at the same time minimizing the adverse effects on normal tissues. It was soon learned that radiation was most effective against microscopic and small volumes of cancer and that surgery was more effective against larger volumes. Ways were sought to combine surgery with radiation to achieve maximum benefit. Radical operation followed by postoperative irradiation became recognized as providing the best chance for cure or long-term control of locally or regionally confined breast cancer. Yet, many women refused such disfiguring and psychologically damaging radical surgery while others were medically unfit for such a major operation. Thus, many women, for a variety of reasons, were treated with lesser surgery.

7.2.2   Early Studies

In 1954, Mustakallio reported a series of women from Finland with cancer confined to the breast with no enlarged glands detectable in the axilla that he had treated with minimal surgery (in which essentially just the cancer lump was removed) and irradiation. His study was a personal case series consisting of 127 women who had been followed post-treatment for at least 5 years. He simply described what he had done and what the outcome was, and made no attempt at statistical analysis. Of the 127 women, 107 (84%) lived 5 years or longer, and 20 had died. Fourteen of the deaths (11%) were from metastases and 6 (5%) were from other diseases. He observed that some women die from metastases later than 5 years and thus observation beyond 5 years is desirable. In his series, 18 patients were followed more than 10 years, and 13 (72%) lived more than 10 years. He estimated that about one–third of all cases of breast cancer could be treated by this method. Although he conceded that 127 patients followed for 5 years was "still too limited a number to allow of a final evaluation of this method of treatment," he went on to conclude that he "may be justified in asserting, however, that the method of treatment in which the breast is saved has passed its experimental stage and can now well be recommended for more general use" (Mustakallio 1954, 26).

Mustakallio's study contains no comparison group, thus he seems to be implicitly using as a comparison similar studies of radical surgery and radiation. Adair (1943) published one such study from Memorial Hospital in New York. During the 22 years prior to publication, over 7000 women with breast cancer were seen at the hospital, of which over 3000 were deemed operable and potentially curable. They were treated by a variety of physicians using evolving techniques of surgery and radiation, but were grouped for study according to whether they received surgery only, surgery followed by radiation therapy, radiation therapy followed by surgery, or irradiation alone. The study was a presentation of what was done and the outcome, and no attempt at statistical analysis was made. The percent of women in each group surviving 5 years was calculated. Since by this time it was known that the presence of axillary node involvement by cancer (positive nodes) in a woman that otherwise had the disease confined to the breast conferred a worse prognosis than if the nodes were negative, the patient groups were further divided into those with nodes positive or negative. In the group that had radical surgery followed by radiation with nodes negative, the group that seems most suitable as a comparison group to Mustakallio's series, the 5-year survival rate was 77%.

Other prognostic factors came to be identified, which when taken into account influenced results. Williams et al. (1953) reported a series of over 1000 cases of breast cancer from St. Bartholomew's Hospital in London that was treated in the 1930s. The cases were divided into stages, with stage I being no enlarged glands could be palpated in the axilla, and stage II where enlarged glands could be palpated. Age was recognized as a prognostic factor and was taken into account and adjusted for. For example, among their stage I cases that were 65 years of age or older, the crude 10-year survival rate was 22%, but when age adjusted it was 49%. These authors found that the age adjusted 5-year survival rate for stage I cases undergoing simple surgery and radiation was 76% and for radical surgery and radiation was 72%. They concluded, "...that where efficient radiotherapy is available radical mastectomy should be abandoned in favor of conservative surgery" (Williams et al. 1953, 796).

Based on reports such as these, some cancer centers began to use limited surgery and radiation therapy to treat early breast cancer, and by the 1970s reports had begun to appear of the results of the more conservative approach. Cope et al. (1976) from the Harvard Medical School reported their experience with 131 women with early breast cancer that had been advised to have a radical mastectomy but refused, but did accept more limited surgery and irradiation. Results were presented according to stage. They concluded that the ". . .survival rates of those treated by limited excision and primary irradiation compare favorably with those of patients treated by radical mastectomy" (Cope et al. 1976, 405).

Spitalier et al. (1986) reported a series of 1133 cases of operable breast cancer treated between 1961 and 1979 by limited surgery and radiation therapy at the Marseille Cancer Institute in France. Follow up ranged from 5 to 23 years. When stage I and II were combined, the percent of patients that were alive and well at 5, 10, and 15 years was 86%, 80%, and 62%, respectively. It was concluded that the conservative approach yielded survival rates equivalent to those achieved by primary radical surgery, and allowed the majority of patients to retain their breasts in an esthetically acceptable condition. In an accompanying commentary from the surgical community on the results of this series, Ketcham (1986, 1019) stated that, "Even today with the literature demonstrating the favorable effects of conservative surgery there remains a large segment of general surgeons who refuse to treat a patient unless they will accept a modified mastectomy." Although he noted also that an increasing number of surgeons were performing more limited surgery, he also opined that, ". . .while one weighs the value of retrospective studies, we must be very analytically critical of some of their conclusions until randomization studies support those conclusions" (Ketcham 1986, 1020).

Haffty et al. (1989) from the Yale University Medical Center reported their 20-year experience in treating 281 women. The paper described their experience and patient outcomes. No statistical analysis was presented. The study was not controlled for prognostic factors, but data were presented separately for stage I and stage II. Patients were staged according to the more modern American Joint Committee/Union Internationale Contre le Cancer system. In stage I, the 5-year survival was 91%, and for stage II it was 68%. They concluded that the treatment of early – stage breast cancer conservatively with limited resection and radiation therapy was a viable alternative to mastectomy.

Comparison of results from similarly conducted studies from different institutions is made difficult by the numerous sources of possible bias, most often from poor control or no consideration of prognostic factors. Even when the "same" prognostic factors are recorded, they may be different. For example, stage I breast cancer is cancer confined to the breast, but is defined differently in the study by Williams et al. (1953), which allows for ulceration of the overlying skin by tumor erosion. Such ulceration would place the case in stage III in the more modern staging system employed by Haffty et al. (1989).

7.2.3 Controlled Trials in Clinical Medical Science Revisited

As has been noted, controlled clinical trials in clinical medical science usually involve two groups of subjects, all of which are suffering from some medical condition; one of the groups, the *test* group, is administered the experimental therapy, while the other, the *control* group, is not; the progress of each group is then monitored over a period. In some trials, there may be more than one experimental group to be compared with controls.

In constituting the control group it is desired that they be as comparable as possible to the test group in regard to the causally relevant (prognostic) factors that could influence the outcome of the study and any inference as to the effectiveness of the experimental therapy. By balancing all of these factors as evenly as possible between the two groups, and leaving the experimental therapy as the only causally active variable, it becomes possible to infer that differences in outcome between the two groups are related to the therapy. This process is referred to as *eliminative induction* (Howson and Urbach 2006, 184).

The question that arises is how to first identify and then evenly distribute these prognostic variables. Fisher (1947) noted that not only are there the *known* prognostic factors, but also a possibly long list of *unknown* factors that could influence results. Fisher developed the processes of *control* and *randomization* to deal with known and unknown prognostic factors, respectively.

A prognostic factor has been "controlled for" in a trial when that factor is equally distributed between test and control groups. For example, if age were a factor, then the groups would have similar age structures. Randomization is designed to evenly distribute unknown factors, although in most cases it should also evenly distribute the known factors as well. Subjects are randomly assigned to treatment or control groups, thus avoiding selection factors that might bias results.

In a clinical trial, how does one control for prognostic factors? One method is to use patients recently treated using some "standard" or accepted form of therapy, a form of "historical" controls. Thus when a new treatment is introduced, the patients receiving the new therapy may be compared to such a historical control group assembled in such a way as to be as comparable as possible to known prognostic factors. Another method is to randomly assign patients to either the experimental therapy or to a control therapy. The control therapy could be some "standard" or accepted therapy, or no therapy at all, such as with a placebo.

7.2.4 RCTs in Early Breast Cancer

In modern medical practice as well as EBM, there is the widespread belief that the "best" evidence on which to base treatment recommendations comes from RCTs. The following statement is typical: "The critical issue regarding primary radiation therapy [for women with early breast cancer] is whether it yields survival equal to mastectomy. This issue can only be answered by randomized prospective trials in which the treatment arms are well balanced in terms of prognostic features" (Recht et al. 1986, 437). Based on uncontrolled or partially controlled studies such as the ones we have already considered, several RCTs comparing minimal breast surgery to traditional mastectomy were undertaken in the U.S. and abroad. In minimal breast surgery, the tumor is removed (variously called lumpectomy, partial mastectomy, segmental mastectomy, quadrantectomy, etc.) and the breast is preserved, usually followed by breast irradiation. I will discuss three of these trials and their results, which are typical.

One RCT was carried out in North America by the National Surgical Adjuvant Breast Project (NSABP) to compare traditional mastectomy with segmental mastectomy in the treatment of early breast cancer (Fisher et al. 1989). Eighty-nine institutions in the U.S. and Canada enrolled a total of 1843 women into the trial, and they were randomly allocated to three groups: total mastectomy, segmental mastectomy, and segmental mastectomy followed by irradiation. The patients were followed for 8 years and results were assessed with standard (classical, frequentist) statistical techniques. Major end points were disease-free survival (alive without evidence of disease), distant disease-free survival (alive without evidence of disease outside the locoregional area), and overall survival (alive regardless of disease status). Ninety percent of the women treated with breast irradiation after lumpectomy remained free of ipsilateral (on the same side) breast tumor, compared with 61% of those not treated with irradiation after lumpectomy ($p < 0.001$). Lumpectomy with or without irradiation of the breast resulted in rates of disease-free survival ($58 \pm 2.6\%$), distant disease-free survival ($65 \pm 2.6\%$) and overall survival ($71 \pm 2.6\%$) that were not significantly different from those observed after total mastectomy ($54 \pm 2.4\%$, $62 \pm 2.3\%$, and $71 \pm 2.4\%$, respectively). There was no significant difference in the rates of distant disease-free survival ($p = 0.2$) or survival ($p = 0.3$) among the women who underwent lumpectomy (with or without irradiation), despite the greater incidence of recurrence in the ipsilateral breast in those who received no irradiation. The authors concluded that the study results support the use of lumpectomy in patients with stages I or II breast cancer, and that irradiation reduces the probability of local recurrence of tumor in patients treated with lumpectomy.

Another RCT in the U.S. comparing conservative treatment with mastectomy was that performed at the National Cancer Institute (Jacobson et al. 1995). Between 1979 and 1987, the Institute conducted a single – institution trial comparing lumpectomy, axillary dissection, and radiation with mastectomy and axillary dissection in stage I and II breast cancer. A total of 247 patients were randomized to modified radical mastectomy (which includes axillary dissection) or to lumpectomy, axillary dissection, and radiation therapy. Randomized patients were followed for a median of over 10 years. At 10 years, overall survival was 75% for patients assigned to mastectomy and 77% for patients assigned to lumpectomy plus irradiation ($p = 0.89$). Disease-free survival at 10 years was 69% for the patients assigned to mastectomy and 72% for the patients assigned to lumpectomy plus irradiation ($p = 0.93$). The rate of locoregional recurrence at 10 years was 10% after mastectomy and 5% after lumpectomy plus irradiation (p = 0.17). The authors concluded that breast conservation therapy with lumpectomy and radiation in stage I and II breast cancer offers results at 10 years that are equivalent to those with mastectomy.

An RCT comparing Halsted radical mastectomy with quadrantectomy, axillary dissection, and radiotherapy in patients with early breast cancer (tumor less than two centimeters in size) with no palpable axillary disease was conducted between 1973 and 1980 at the National Cancer Institute in Milan, Italy (Veronesi et al. 1981). Randomized patients numbered 701, with 349 receiving Halsted mastectomy and 352 receiving quadrantectomy. Survival curves showed no difference between the two groups in disease-free survival (p = 0.54) or overall survival (p = 0.88). The authors concluded that "... radical mastectomy appears to involve unnecessary mutilation in patients with carcinoma of the breast measuring less than 2 cm and without palpable axillary nodes" (Veronesi et al. 1981, 11).

7.3   Is Randomization Necessary?

The view that randomization of subjects to test and control groups is the best method currently available for eliminating potential bias in controlled clinical trials has been challenged from within both the philosophical and medical communities. Howson and Urbach (2006), for example, in arguing in favor of Bayesian over classical, frequentist statistical approaches to clinical trial analysis, offer several criticisms of randomization. Worrall (2007) has also argued that randomization is not necessarily to be preferred. Similarly, criticisms are found within the medical community, such as those advanced by Freireich and Gehan (1979).

Howson and Urbach challenge the notion that randomization is absolutely essential and state that "... the problem of nuisance variables in trials cannot be solved by its means," and maintain that the essential feature of a trial that permits a satisfactory inference of causal efficacy of a treatment "...is the presence of adequate controls" (2006, 187). Freireich and Gehan (1979, 282) similarly state: "The most important element to control is the comparability of patient populations receiving different treatments." Howson and Urbach examine two main arguments for randomization found in the literature: (1) that of classical statisticians for which significance tests are central to inference that such tests are valid only if experiments were randomized, and (2) an eliminative-inductive defense.

In evaluating (1), Howson and Urbach emphasize that the population subjects are drawn from for random sampling is often poorly defined. They consider the randomization step itself, i.e., that just those patients randomized could be considered the population. They find this unsatisfactory, however, since one aim of the trial is to extend results to other populations, for example people in faraway places or those presently healthy or as yet unborn who will develop the disease in the future. They state that, "...in no trial can random samples be drawn from hypothetical populations of notional people" (2006, 190). They further state "...the path from "representative sample" to "general body of patients" – two vague notions – cannot be explored via significance tests and is left uncharted; yet unless that path is mapped out, the randomized clinical trial can have nothing at all to say on the central issue that it is meant to address" (2006, 191).

Howson and Urbach also cite Kendall and Stuart as implying that the randomization process is subjective: "A substantial part of the skill of the experimenter lies in the choice of factors to be randomized out [i.e., distributed at random] of the experiment. If he is careful, he will randomize out all the factors which are suspected to be causally important. . .every experimenter necessarily neglects some conceivably causal factors. . ." (2006, 193). They then go on to say that this subjectivity is at odds with the classical methodology underlying Fisher's argument.

In (2), in the eliminativist-inductive defense, it is argued that randomization tends to balance prognostic factors whether or not they are known. As Howson and Urbach state, it is essentially that ". . .although randomization does not give a complete assurance that the experimental groups will be balanced, it makes such an outcome very probable, and the larger the groups the greater the probability" (2006, 195). They think that this is mistaken unless significantly modified, and the modified position, while compatible with Bayesian thought, is inimical to classical inferential theory.

Howson and Urbach consider that the number of unknown prognostic factors could in fact be quite large. In addition, they raise the possibility that there could also be factors not only related to the subjects but to the treatment, for example, impure drug or different treatment environments. They conclude that the objective probability could range anywhere between zero and one. They reject the strong claim for randomization, that it *guarantees* the groups to be balanced, but also find fault with the more popular, weaker claim that it *probably* or *tends to* balance the groups, stating that this weaker claim ". . . cannot exploit eliminative induction. For, the premise that the experimental groups were *probably* balanced does not imply that differences that arise in the clinical trial were *probably* due to the experimental treatment, unless Bayes's theorem were brought to bear, but that would require the input of prior probabilities and the abandonment of the classical approach" (2006, 197).

Howson and Urbach concede that randomization is not necessarily harmful, nor claim that it is never useful, but deny that it is absolutely necessary. They consider it undesirable that a requirement for randomization makes the use of historical controls illegitimate, and state: "The control group could be formed from past records and the new treatment applied to a fresh set of patients who have been carefully matched with those in the artificially constructed control group (or historical control)" (2006, 202). Freireich and Gehan also state: ". . .an unfortunate consequence is that the unqualified acceptance of the randomized clinical trial has often led to the rejection of the validity of historical data" (1979, 278). They also argue that data from an RCT is of no better quality than that generated by a test group compared with historical controls (1979, 295). Howson and Urbach consider ethical issues in randomization, stating:

A new treatment, which is deemed worth the trouble and expense of an investigation, has often recommended itself in extensive pilot studies and in informal observations as having a reasonable chance of improving the condition of patients and of performing better than established treatments. But if there were evidence that a patient would suffer less with the new therapy than with the old, it would surely be unethical to expose randomly selected sufferers to the established and apparently or probably inferior treatment. Yet this is just what the theory of randomization insists upon.

No such ethical problem arises when patients receiving the new treatment are compared with a matched set of patients who have already been treated under the old regime (2006, 203).

Freireich and Gehan also address ethical concerns with the RCT. They state:

...the ethical basis of the randomized clinical trial is questionable because therapeutic research depends upon the investigation of new treatments that have a greater potential for benefit than for risk compared with a standard treatment, and this is a circumstance in which it would be unethical to randomize patients (1979, 294).

Howson and Urbach believe that a Bayesian approach to clinical trials is preferable to classical approaches involving randomization, and state: "...despite the weight of opinion that regards it as a *sine qua non,* the randomizing of treatments in a trial does not do the job expected of it and, moreover, that in the medical context, it can be unnecessary, inconvenient, and unethical" (2006, 254). They recognize the need for controls in clinical trials, however, as necessary to estimate how test patients would have fared without the therapy. They believe that Bayes' theorem provides a rational basis for selecting controls, and extol the virtues of historical controls. They, like Freireich and Gehan, have no problem with the admissibility of such controls, apparently matched on the known prognostic factors. Freireich and Gehan as well as Howson and Urbach point to the practical advantages of using historical controls, like the need to treat fewer subjects and in general being less expensive to conduct. Historical comparison groups don't expose patients to ineffective placebos, "... considerations that address natural ethical concerns, and mitigate the reluctance commonly found amongst patients to participate in trials" (Howson and Urbach 2006, 261). They acknowledge, however, that such historically controlled trials aren't easy to set up, relying often on thorough medical records that are more detailed and accessible than those routinely available.

Howson and Urbach believe that the Bayesian approach to clinical trials is superior to the classical approach. They state:

Bayes's theorem supplies coherent and intuitive guidelines for clinical and similar trials, which contrast significantly with classical ones. One striking difference between the two approaches is that the second simply takes the need for controls for granted, while the first explains the need and, moreover, distinguishes in a plausible way between factors that have to be controlled and those that do not (2006, 262).

They continue:

We have shown that frequentist tools do not solve *any* problems. The conclusions they license ... have no inductive significance whatever. True, one can often draw frequentist conclusions "easily", but this is of no account and does not render them scientifically meaningful. True, many scientists feel "comfortable" with frequentist results, but this, we suggest, is because they are misinterpreting them... (2006, 264).

Worrall also argues that Fisher's belief that randomization is required to justify the use of significance tests is unnecessary, and that multiple other factors could exist which might explain the outcome of an RCT (2007, 996–1001). He also agrees with Howson and Urbach that randomization might not succeed in evenly distributing confounding factors, and that classical, frequentist statistics makes no intuitive sense because it has no formal method for assigning probabilities to hypotheses (2007, 1001–1008). He does, however, grant that preventing selection bias is a "cast-iron argument for randomization" (2007, 1009). He adds that randomization is only a means to an end, and not an end in itself. The aim is to take away from experimenters control over which arm of the trial patients are assigned to, and that randomization is just one way of achieving this.

Although they do not champion a Bayesian approach to data analysis, Freireich and Gehan also criticize what they regard as an over – reliance on statistics. For example, they point out that a statistically significant difference can always be shown between any two treatments if the samples are large enough, and state, "One of the difficulties in the application of the scientific method to clinical studies is the impossibility of proving a null hypothesis" (1979, 286).

7.4 A Defense of Randomization

Critics of randomization minimize its importance in clinical trials and emphasize the value of historical controls in assessing the effectiveness of novel therapies. They also draw attention to greater expense and inconvenience, as well as negative ethical aspects of such RCTs. Howson and Urbach (2006) criticize the frequentist interpretation of parameter estimates as nearly useless since subjects are not drawn randomly from a defined population.

I believe that such critics have underestimated the importance of randomization, and that it is quite often absolutely necessary, particularly to help settle controversies over competing therapies for major illnesses. Their belief that historical controls can suffice to the extent that they imply is unjustified. Their comments on negative ethical aspects are mostly one-sided and tell only half the story. Howson and Urbach (2006) have not told us under what circumstances it is necessary to randomize, only that it is not *absolutely* necessary, and much of their case seems to rest on a confusing intermingling of what have been called *statistical* and *scientific* inferences.

By *statistical* inference is meant inferring from a noncontroversial statistical point of view arising from the assumptions made and application of the probability axioms. Examples would be the probability of 0.5 of getting heads on the toss of a fair coin, or estimating the mean height θ of a large population of people, whose standard deviation σ is known, by the sample mean $\bar{x}$. As Howson and Urbach note, because the distribution is essentially normal, it follows that, with a probability of 0.95, $-1.96\,\sigma_n \leq \theta - \bar{x} \leq 1.96\,\sigma_n$ (2006, 169–170). By *scientific* inference is meant an inference that may, though not necessarily, involve a statistical inference but is a generalization not justified by strictly statistical considerations. Examples would be inferring that, *ceteris paribus,* women in Denmark will respond similarly to a treatment for breast cancer administered to women in New York, or that children everywhere would respond similarly to children in Rio de Janeiro successfully vaccinated against measles.

First, what of the arguments by Worrall (2007) and Howson and Urbach (2006) against Fisher's notion that random samples are necessary to underwrite significance tests? This perhaps appears to be a straw man: Significance tests are used not only in RCTs but are also frequently used in studies where no randomization has been carried out, such as in the analysis of cohort studies. And Papineau (1994), a defender of randomization, also concedes that random samples in RCTs are unnecessary, but also argues, as I do below, that it misses the point of randomization in RCTs.

What of the claim by Worrall (2007) and Howson and Urbach (2006) that the problem of nuisance variables can't be solved by randomization? These "nuisance variables" are the prognostic factors that can't be controlled for because they are unknown. But if they are unknown to frequentists, they are also unknown to Bayesians. Both frequentists and Bayesians would agree that control of *known* prognostic factors is essential, but also that no threat to validity by bias from *unknown* prognostic factors is necessary. Frequentists offer randomization as a solution, albeit an imperfect one. What is the Bayesian alternative? None is on offer. But interestingly, Howson and Urbach also state that randomization may "sometimes be useful" to better balance groups, apparently by helping to eliminate selection bias (2006, 259). And, Worrall agrees that randomization is a valid method to control selection bias (2007, 1008–1009).

It is comforting to know about selection bias, but if we need to randomize to eliminate a bias like this that we know about, what about the possible biases that we *don't* know about? The main reason for random allocation to begin with is to balance unknown prognostic factors. And, selection bias is not limited to selection by experimenters of which group to assign a patient to, as these critics seem to imply.

Selection bias is one of several types of biases, which along with confounding, is a threat to study validity, and thus accuracy. As I have argued, it is particularly problematic with observational studies such as cohort and case-control studies, in which study groups must somehow be selected. But the problem is the same: ensuring to the extent possible that groups are comparable except for the factor (s) under study. How should patients in an RCT be assigned to study groups? Should patients themselves select which group they will enter? Self-selection is a known threat to validity, as illustrated in the Smoky nuclear study (Caldwell et al. 1980) discussed in Chap. 5, so this method would be unsatisfactory. The problem is that if patients are not assigned randomly, then some *nonrandom* method of assignment must be put in place. This implies that some *criteria* for assignment would be required. These criteria presumably would specify how patients would be *selected* for assignment to the study groups. Thus, possible selection factors might be introduced that could bias study results, exactly what randomization seeks to avoid. Critics like Worrall (2007) and Howson and Urbach (2006) have offered no plausible alternative to the RCT to control selection bias, and I believe the reason is that there may well not be one.

Many critics seem to believe that a new therapy can be evaluated by forming a test group and comparing it with a control group already treated with some comparison therapy and carefully matched on known prognostic factors without the need for randomization. But the use of historical controls is fraught with pitfalls (Byar et al. 1976). For example, in RCTs, great attention is usually paid to the setting up of standardized definitions, not necessarily the case in historical settings. Also, definitions may change over time. More sensitive diagnostic procedures may systematically detect earlier, thus less severe states. The same names may be used to describe these new states, thus leading an investigator to think the states are comparable. Changes in general supportive care may also occur over time, affecting prognosis in countless subtle ways. Treatment techniques often vary considerably in nonrandomized settings. For example, the "same" surgical operation may be performed with several variations, and techniques may improve over time. Already noted above in the studies by Haffty et al. (1989) and Williams et al. (1953) is how the "same" stage of disease can actually be quite different. The use of laboratory and other diagnostic tests may vary greatly depending on personal preferences by practitioners. In RCTs, there is standardization and direct comparability between test and control subjects that is designed to remove these potential obstacles. Test and control subjects are derived from the same group of eligible patients, and are *concurrently* randomized to the study groups. Finally, historical controls even in principle can only control for *known* prognostic factors. Green et al. (1997, 143–150) describe several examples of erroneous conclusions when historical controls have been relied on in nonrandomized studies, and Mayo (2005, 97) also provides the example of misleading inferences from the nonrandomized studies of hormone replacement therapy that weren't appreciated until the later RCTs. And, not to be forgotten is the observation from the Gastrointestinal Tumor Study Group RCT described earlier, in which the authors stated: "... the importance of a no adjuvant therapy control arm cannot be overemphasized. Our results for this cohort of patients are much better than would be suggested by historical controls" (Thomas and Lindblad 1988, 250). Such observations are by no means unusual among those conducting RCTs.

Freireich and Gehan (1979) as well as Howson and Urbach (2006) apparently believe that if some evidence exists that a new therapy is better than established therapy, an RCT to better establish this is unethical. For example, Howson and Urbach state: "But if there were some evidence that a patient would suffer less with the new therapy than with the old, it would surely be unethical to expose randomly selected sufferers to the established and apparently or probably inferior treatment" (2006, 203). But, I think this belief is mistaken.

The problem is that "some" evidence usually isn't enough to establish a new form of therapy in place of an old. If the illness is serious, like cancer for example, RCTs are usually necessary to convince practitioners to adopt the new therapy or for organizations like the American Society for Radiation Oncology or the NCCN to alter treatment guidelines.

The dominant ethical justification for randomization is *clinical equipoise* (Freedman 1987). It is based on the recognition that the purpose of an RCT is *social*, to change standards of medical practice. Clinical equipoise exists when there is uncertainty or disagreement among the *expert medical community* about which intervention is better. As Freedman notes, physicians "must simply recognize that their less-favored treatment is preferred by colleagues whom they consider to be responsible and competent" (1987, 144). It is this absence of consensus among medical experts about what is the best treatment that ethically justifies the RCT.

This is not to say that insistence on performing an RCT when other arguably good evidence exists of therapeutic effect never raises ethical concerns. I have argued that we have done a sufficient number of studies to confirm a hypothesis when additional study is believed unlikely to change the weight of evidence for that hypothesis. I address these issues further in Chap. 8.

Howson and Urbach claim that the randomization process is necessarily subjective, citing Kendall and Stuart as noted above. They state: "What Kendall and Stuart demonstrated is that randomization has to be confined to factors that the experimental designers judge to be of importance, and that this judgment is necessarily a personal one, which cannot be based solely on objective considerations" (2006, 194).

The experiment referred to by Howson and Urbach and discussed by Kendall et al. (1983, 135–137) concerned the effects of blood alcohol levels on reaction times among male automobile drivers. The drivers were randomly assigned to various doses of alcohol and their reaction times and blood alcohol levels were measured after a fixed interval. They made the point that if the time of day that the experiment was conducted interacted with reaction times and blood alcohol levels, then it could be studied as a factor in a regression analysis after the experiment whether or not it had been randomized out (distributed at random). But if the time of day that the tests were done had been fixed, for example at 6 p.m., then it couldn't be studied. What variables to fix or randomize out are subjective and left to the investigator.

This is fine as far as it goes, but it is hard to see how this is a criticism of an RCT. We want to eliminate bias and confounding, which may result from the uneven distribution of some prognostic factor(s) between experimental groups. If a factor is fixed, it will not produce bias. If it is randomized out, we expect it to be distributed randomly (and evenly) between the groups, thus avoiding bias. This example seems to not go through as a criticism of randomization in an RCT.[2]

---

[2]The problem being alluded to here is very real, but it is a general one: in conducting any scientific study, it is possible that one or more factors, unknown to the investigator(s), may be present and produce a distortion in one or more outcome measures. "Fixing" a factor usually is done purposely to minimize or eliminate a bias, for example, age-matching. The problem, *per se*, is not confined to randomization or RCTs, hence is not an argument against these methods. It is, of course, one reason why more than one study is highly desirable in some cases.

Howson and Urbach criticize clinical trials from the standpoint of what they regard as the inapplicability of classical statistical techniques to samples not drawn randomly from a well-defined population. As noted above, they state that, "... in no trial can random samples be drawn from hypothetical populations of notional people" (2006, 190). They are correct, but it is beside the point. Usually no effort is made to draw patients randomly from some defined "population," hypothetical or real. Of course, patients must meet eligibility criteria. Consider as an example the RCT carried out by the NSABP discussed earlier (Fisher et al. 1989). The important point in this trial (and the other RCTs in early breast cancer) is that the only "random" aspect was the method of allocation of the patients to the experimental groups. When an eligible patient appeared at a participating institution, she was offered the opportunity to participate. If she agreed, she was randomly allocated to one of the three experimental groups by the central NSABP office. No effort to form a "representative sample" was made. So it is difficult to assess their statement that "... the path from "representative sample" to "general body of patients" – two vague notions – cannot be explored via significance tests and is left uncharted; yet unless that path is mapped out, the randomized clinical trial can have nothing at all to say on the central issue that it is meant to address" (2006, 191).

But what is the "central issue" to which they refer? For the NSABP and other similar randomized trials, the research question is whether the less disfiguring partial breast removal combined with radiation therapy can replace whole breast removal without compromising expected outcomes. "Representative samples" have nothing to do with it. It is true that significance tests are applied to the groups in the trial, but any inference from the women in the trial to a woman elsewhere is a scientific inference: it is that a woman with similar characteristics to the women in the trial who is treated similarly can reasonably expect a similar outcome. It is *not* a statistical inference that is hampered by the lack of a formal procedure for assigning probabilities to hypotheses in classical (frequentist) statistics.

Howson and Urbach also maintain that eliminative induction cannot be used to justify the weaker claim that randomization *tends to* balance the groups, stating "...the premise that the experimental groups were *probably* balanced does not imply that differences that arise in the clinical trial were *probably* due to the experimental treatment, unless Bayes's theorem were brought to bear, but that would require...the abandonment of the classical approach" (2006, 197). They grant, however, that the strong claim for randomization, which is that it *guarantees* that the groups are balanced, if true, "...then the conditions for an eliminative induction would be met, so that whatever differences arose between the groups in the clinical trial could be infallibly attributed to the trial treatment" (2006, 196–197).

Let's look more closely at their argument: If we *knew* that the groups were balanced (the strong claim), then we would *know* (infallibly) that differences could be attributed to the trial treatment. This is of the form: If A, then B. In probabilistic terms: If $P(A) = 1$, then $P(B) = 1$. But then they say we can't say that if *probably* A, then *probably* B unless we invoke Bayes' theorem, assign priors, etc., and abandon classical statistical methods. Is this true?

I would argue no, for the following reason, and here I will assume that *probably* means a probability that is greater than 0.5 but less than 1: We have strong evidence from computer modeling and computer – based simulations that random assignments like those carried out in RCTs distribute factors evenly among experimental groups with a high probability, which increases as the number of patients increases. It is *not* a subjective assessment by investigators.

Consider, for example, an RCT testing some treatment A against a treatment B. Suppose results favor treatment A (p < 0.05). Thus, the probability that this result or a result more extreme favoring treatment A occurred simply by chance is less than .05. If I have good reason to believe (that is, the probability is high) that the groups were balanced with respect to possible confounders, and that other potentially biasing factors have been controlled, then it would seem eminently plausible to conclude that *probably* treatment A was responsible for the result. Or, stated differently, probably treatment A *caused* the result (Papineau 1994; Cartwright 2007). No resort to Bayesian reasoning is required.

Now of course one *can* use a Bayesian approach to the analysis of the results of an RCT. But this is not the inevitable result of whether or not we can use eliminative induction, for Bayesians are no better off here than frequentists. We can never be *certain* that the groups were balanced (the strong claim), thus we can never *know* (infallibly) that differences could be attributed to the trial treatment.[3] But science, which includes clinical medical science, as most surely recognize, isn't about certainty. It is about, among other things, eliminating possible sources of error to the extent possible before making inferences based on the results of the trial. In the analysis of most RCTs, standard, classical statistical techniques are used to study differences in the endpoints (e.g., survival curves) generated describing the groups. If they are so different that the probability of observing a difference that extreme or more extreme is very low (say, p < 0.05), then we are to that extent confident that the test treatment is causally active. By randomizing we have done everything we know how to do to reduce bias from uneven distribution of unknown prognostic factors. It is difficult to see how we could be *more* or even *as* confident in the results of a Bayesian analysis in which only the *known* prognostic factors have been controlled for, and randomization has been voluntarily dismissed as unnecessary.

Critics often overlook how at least many, if not most, RCTs are actually conducted. It must be remembered that randomization tends to balance both known and unknown factors, and thus it is not surprising that all known prognostic factors are often not controlled for before randomization under the expectation that they will be evenly distributed by the randomization process itself. Thus, in the NSABP trial discussed earlier (Fisher et al. 1989), all eligible patients were randomized without prior control for all known prognostic factors. In Table 3 from the 5-year results of this trial (Fisher et al. 1985), it was shown that several known

---

[3]Howson and Urbach (2006, 197) state (and I believe most would agree) that the strong claim is indefensible.

prognostic factors were evenly distributed among the experimental groups, adding confidence that unknown prognostic factors were also evenly distributed.

So, after randomization, if we can show (that is, we have *evidence*) that known prognostic factors have been evenly distributed, it seems that we can be more justified that the unknown prognostic factors have likewise been evenly distributed than in the Bayesian approach as outlined by Howson and Urbach (2006), in which we have no assurance whatever that unknown factors have been evenly balanced. Thus, Simon (2008, 582) seems quite justified in asserting that, "...some people believe that Bayes' theorem is somehow a substitute for randomization. In fact, however, randomization is just as important for the validity of Bayesian methods as for frequentist methods."

7.5 Analysis of Evidence in Studies of Treatment for Early Breast Cancer

Radical surgery originally became the accepted treatment for locoregional breast cancer because there were no alternatives. With the introduction of radiation therapy, it became possible to consider less disfiguring and mutilating surgery. If the lesser approach resulted in no lower probability of survival or of other outcome measures, the lesser treatment would become a viable, and in general, preferred alternative.

But to adopt such a new treatment approach, particularly for a serious, and often fatal, disease like breast cancer, what is required is *convincing evidence* that the new approach will not reduce a woman's chances of surviving the disease. Thus the *hypothesis* being entertained, in this case, is a *null* hypothesis, that of no difference in outcome.

I suggest that at least three "levels" of such evidence can be identified from the historical evolution of the studies assessing the conservative approach. The first level, Level I, is the least convincing, and comes from studies that simply adopt lesser surgery and radiation therapy in a series of patients and then compare it to some series of patients treated by radical surgery. Simple measures, like the percent of women alive at 5 or 10 years, are presented. In this level, no effort is made to control for prognostic factors that could bias results. An example is the study by Mustakallio (1954). A second level of evidence, Level II, "better" and more convincing than the first, takes into account prognostic factors and attempts to "control" for them in some way. Studies in this level are more heterogeneous than in the first level and include studies that control for only one or two prognostic factors as well as studies that try to control several or even all known prognostic factors. Examples are the studies by Williams et al. (1953) and Cope et al. (1976). The third level, Level III, I contend provides the "best" or most convincing evidence that clinical medical research can aspire to. It is that provided by RCTs (which also lead to their derivatives, such as systematic reviews or meta-analyses of RCTs) that control to the best of our ability *all* prognostic factors, known and unknown.

Analyses of the survival outcomes in Level I and II studies were quantitative, and simple descriptive statistics were used, including simple adjustment techniques in some Level II studies. Level III studies employed classical, frequentist statistical methods and were designed to pose what Mayo (2005) would consider a severe test, and what I am calling an accurate study or test. More specifically, the analyses used in the type of RCTs directed at detecting survival differences in the populations tested make use of sophisticated statistical techniques, including those that utilize life table (Cutler and Ederer 1958) and log-rank methods (Mantel 1966). These methods pose a much more sophisticated analysis than would a simple 5 or 10 year survival percentage, even if the study was an RCT in which all prognostic factors had been evenly balanced between the two groups. This can be illustrated by a simple (albeit extreme and implausible) example. Suppose in a hypothetical RCT that 1000 women were given Treatment A and 1000 women were given Treatment B, and each group was followed for 5 years. Assume further that 900 women given Treatment A died in the first year and 100 survived until the end of year five, and that all 1000 women given Treatment B survived until the beginning of year five, and during that year 900 died with only 100 surviving until the end of year five. The 5 year survival rate for both Treatments A and B is 10%, but clearly the treatments have produced quite different results, with many more women given Treatment B surviving many more years overall. These differences are captured and quantified in modern survival analyses.

In none of the RCTs was evidence provided that one of the treatment arms was superior. Many surgeons believed that any form of therapy short of radical removal had to be inferior, and certainly the Level I and II studies could be (and were) criticized as possibly biased due to poor or no control of prognostic factors. So, it would seem, the RCTs have provided the most convincing evidence possible of the truth of the null hypothesis of no difference between mastectomy and the conservative approach. Or are we entitled to that claim? Certainly, most in the health care field would say that we are. But how should we react to Mayo's statement that "... taking no evidence against the null as evidence for it is a well-known fallacy" (2005, 111)?

Mayo does not fully develop this notion here, however. Certainly, if we fail to reject the null hypothesis in an RCT at the conventional level of $p < 0.05$ with a p-value of, say, 0.06, then we would hardly be justified in claiming that the results of the trial constituted evidence *for* the null hypothesis. In questionable cases, additional studies must be done. However, it may not be necessary to claim evidence *for* the null hypothesis. For if the purpose of these studies is construed as assessing the claim that radical surgery is "better" (in the appropriate sense), then radical surgery has failed to pass the severe (in Mayo's parlance) test imposed by any of the RCTs. In none of the trials was the p-value even remotely close to values that would be considered as even suggestive that radical surgery was superior.

But if the null hypothesis is indeed true, how can we obtain evidence for it? It is not uncommon to see RCTs designed to test the *noninferiority* of some less radical or toxic therapy against a more standard therapy, which is essentially a test of a null hypothesis, such as the study by Muss et al. (2009) comparing standard multi-drug chemotherapy to single agent capecitabine in older women with early breast cancer. But even if we grant that failure to reject the null hypothesis in a single RCT may not be evidence *for* it, could performing several RCTs in which the null hypothesis is not rejected provide such evidence? This would seem to be at least part of the rationale for a meta-analysis of RCTs. Indeed, as Greenhalgh observes, the meta-analysis sits at the pinnacle of the traditional hierarchy of evidence (2010, 44).

Greenhalgh states that: "RCTs are often said to be the gold standard in medical research" (2010, 37). She observes that in nonrandomized controlled clinical trials that the baseline differences between the groups being compared very often are so great as to invalidate any difference ascribed to the intervention (2010, 52–53). In a meta-analysis of RCTs, she notes, results from a number of similarly conducted studies addressing the same research question are pooled, with the objective of obtaining a more accurate estimate of effect. Point estimates and confidence intervals from the various trials are combined to achieve a single point estimate and confidence interval (2010, 121–128).

A meta-analysis of results of nine RCTs comparing mastectomy with breast conservation therapy in the treatment of early breast cancer was performed (Early Breast Cancer Trialists' Collaborative Group 1995). A total of 2423 women were randomized to mastectomy and 2468 to conservative therapy. There was no apparent difference in total mortality ($p = 0.7$), rates of local recurrence, or other outcome measures studied.

Historically the hypothesis that developed was that breast conservation therapy yields results equivalent to mastectomy. The major clinical question for physicians is: Can patients with early breast cancer be told that, to the best of our knowledge, conservation therapy yields survival and other prognostic end results comparable to mastectomy, and is a viable treatment option? If so, then a woman's choice of therapy can be based on other factors. For example, a woman that highly values preservation of her own breast may elect to undergo breast conservation, which requires several weeks of daily radiation treatments, whereas a woman for which it makes little difference may elect mastectomy and forego the requirement for radiation.

So ideally what we seek is evidence for the hypothesis of equivalence. But as Freireich and Gehan (1979, 286) note, we can never "prove" a null hypothesis, since if we make the sample size large enough we can in principle obtain evidence of an effect for one of the treatments at whatever statistical level of significance that we choose. But science, including clinical medical science, is not about certainty or "proving" anything, as I noted earlier. But let us assume that the two forms of therapy are indeed equivalent. Then if we did several very large RCTs such that at some level of statistical significance, say, $p < 0.05$, that each study showed a significant effect for one form of therapy, then we would expect to see about as many studies "positive" for mastectomy as we would for conservation therapy, since the probability of a positive result for either is 0.5. We do not have this, but we do have the results from the meta-analysis of nine RCTs. What we see is that in three of the trials the point estimates are to the left of the vertical "no effect" line of odds ratio 1.0 (Greenhalgh 2010, 123) and six are to the right (the side favoring conservative therapy.) All 95% confidence intervals cross the "no effect" line, so this meta-analysis passes the Chi-squared test for homogeneity (Greenhalgh 2010, 126). These data are consistent with the null hypothesis of no significant difference between treatments. Certainly, there is no evidence that *favors* mastectomy.

Applying the test of significance from the binomial distribution assuming $n = 9$ trials with $x = 6$ binary outcomes, $p = 0.5$ for each outcome, we find that the probability of obtaining this result or a result more extreme (that is, $B \geq 6$) = 0.254. There is no statistical evidence to suggest a departure from the null value. Thus, I argue that we are justified in making the *scientific* inference that there is no apparent difference between the two forms of therapy. From a practical point of view, as clinical scientists, it is reasonable to believe that we have done as much as reasonably possible to rule out the possibility that one form of therapy is superior. This is at the heart of EBM.

Let us briefly review the evolution of studies concerning treatment of early breast cancer. The earliest studies consisted of trying the new approach (conservation therapy instead of traditional mastectomy), and simply reporting the results, without any indication of a comparison group or statistical analysis. I have called such studies Level I, and the study by Mustakallio (1954) is an example. These studies suggested that the new approach may be beneficial, and prompted further study. Level II studies are from institutions that try to identify a comparison group and may try to control for factors that affect outcome, such as age. Some make an attempt at statistical analysis. Examples are the studies by Williams et al. (1953) and Haffty et al. (1989). Such studies cannot control for all biases, particularly those that are unknown. Stage of disease is one such potential bias, and as noted previously, stage was assessed differently in these two examples from Level II. Differences among studies in Level II with poor control of bias make it difficult to form confident conclusions, although they generally supported the hypothesis of essential equivalence between the two therapies.

Level III studies, RCTs, provide the best evidence since maximum effort is directed at strict comparability between treatment and control groups with sophisticated statistical analyses. Three typical RCTs were discussed, all with the interpretation that the therapies were equivalent. This was followed by the meta-analysis of nine RCTs that also could identify no significant difference. Thus, the weight of evidence favoring the hypothesis of the essential equivalence of mastectomy and conservation therapy seems very strong.

7.6   Some Issues in Generalizing the Results of RCTs in Clinical Medical Science

I have argued that generalizing from RCT results in clinical medical science is a scientific inference rather than a statistical inference. The extent to which the results from any RCT or group of RCTs addressing the same issue, for example, some new therapy, can be generalized will depend on numerous factors, not the least being the extent to which patients that might be considered candidates for the new therapy are comparable to trial patients. This is sometimes referred to as *external validity*, and the larger population to which results are to be generalized is sometimes referred to as the *target* population (Cartwright 2007).

Ideal RCTs are one of a group of methods that Cartwright calls "clinchers" (2007, 14). According to Cartwright, it can be proved that if the auxiliary assumptions are true, the methods are applied correctly and the outcomes are true and have the right form, then the hypothesis must be true. They work deductively, a view shared by Howson and Urbach (2006, 197). Another characteristic Cartwright ascribes to RCTs is that they are *self-validating* (2009, 130). Manuals exist for the correct conduct of RCTs to ensure that all of the assumptions are met, including randomization for the equal distribution of confounders, blinding, and other tactics. However, Cartwright views these positive characteristics of RCTs to come at a price: *narrowness of scope*. How do we justify exporting a causal claim from the experimental population to a target population? For Cartwright, external validity for RCTs is hard to justify (2007, 19).

Cartwright considers evidence-based policy, and argues that not only must evidence claims be *credible* (likely to be true), but also that "the full body of evidence should make the conclusion probable, or probable enough given the size of the policy bet" (2009, 128). These concerns go straight to the heart of EBM and the conclusions derived from studies that form the basis for policy recommendations, such as the warning of the U.S. Surgeon General about the hazards of cigarette smoking or the treatment guidelines for women with early breast cancer promulgated by the NCCN. Results from RCTs must be *relevant* to the target population.

Erroneous generalizations from RCTs conducted in one setting to another, different setting may certainly occur, and Cartwright illustrates this with an RCT done in Tennessee that showed that school class size reductions improved reading scores (2009, 131–132). The California class-size reduction program, in which class sizes were reduced in an effort to improve reading scores, used as evidence the well-conducted RCT done in Tennessee. Yet, in California, when class sizes were reduced, reading scores did not go up.

Apparently, significant differences existed between Tennessee and California, and there is a conventional explanation for the unexpected results. California implemented the program over a short period, which created a sudden need for new teachers and new classrooms. Significant numbers of poorly qualified teachers were hired, and the more poorly qualified teachers went to the more disadvantaged schools. Also, classes were held in inappropriate spaces and other educational programs thought to be conducive to learning to read were curtailed due to lack of space. In addition, it was thought likely that the distribution of confounding factors already in place were different in California than in Tennessee.

Roush (2009) also addresses the problem of generalizability (which she calls *transferability*) and notes that it is not peculiar to RCTs but to any kind of study, and that although the studies themselves offer no solution, the problem can be addressed through further studies. Thus, for example, although we could see some specific result in one RCT, we would be considerably more confident if the same or nearly the same result were seen in several RCTs studying the same intervention, since it would be expected that spurious results due to local bias or confounding in one study would not be replicated. This replication of results is what we have in the RCTs of early breast cancer described earlier, which, as noted, adds confidence in the accuracy of the findings.

Roush believes that we should resist the expectation that there will be a single time when the generalizability problem for a study result is solved (2009, 142). Nevertheless, by utilizing common sense, practitioner experience, observational studies and the like, we can continue to uncover additional potentially confounding factors for each new round of RCTs and other studies (2009, 144–145).

Thus, findings from RCTs do not *guarantee* generalizability, and even may not be generalizable at all. How do RCTs in clinical medical science fare? There is, I believe, no simple answer to this question, but on reflection, it is a problem not just for RCTs, but as Roush (2009) notes, also for any type of study in clinical medical science. Indeed, it would seem that the problem permeates nearly all, if not all, of the inductive empirical sciences. It must be rare indeed that scientists are able to study *all* of the individual units, elements, or individuals of the subject or phenomenon that they are investigating. Virtually every study can be considered the study of a sample in which results at least in principle might be generalized to some larger population. This is as true of the physical sciences as it is of the social sciences, the latter of which would include the Tennessee RCT.

Consider, for example, the boiling point of water (H_2O). No doubt water has been found countless times and in countless geographic locations to reliably boil at 100 °C (assuming appropriate purification and under the appropriate experimental conditions). Nevertheless, these are observations on only a miniscule fraction of *all* the water that exists. We confidently treat the statement "Water boils at 100°C" as the statement of a law of nature, or an indisputable fact that is generalizable to the maximal extent. We treat it thusly because we believe water molecules to be *homogeneous* with respect to their physical properties. Providing that purity and the other experimental condition caveats are met, there is every reason to believe that, for example, water from my faucet and a water sample from Lake Tanganyika will both boil at 100 °C without any need for prior special testing of these water sources for verification.

Objects of study in the biological sciences are known to be, in general, not so homogeneous as those in the physical sciences, but at least we are still in the realm of the natural sciences, where we expect a certain order of nature to prevail, as experience has amply shown. Contributing to the relative homogeneity of human beings is the fact that we are of a single species, *Homo sapiens*, and we can reliably expect normal humans to share the same anatomy and physiology, and to respond similarly to medical interventions such as drugs or vaccinations, and to physical stimuli such as scalding water producing burns. Worrall, in his comments on RCTs, has noted that many interventions, such as aspirin for mild headache and appendectomy for acute appendicitis, are very well established even though they were not subjected to an RCT (2007, 986). Surely this observation is just another indication of the relative homogeneity of humans, and that interventions such as these, which are almost too numerous to count, would be equally applicable to patients whether they lived in London or Nairobi, without the need, as in the boiling water example, for special prior testing of patients in these two locations to verify this.

Nevertheless, demonstrably some inhomogeneity exists, such that we cannot be *certain* that all humans will always react exactly in the same way to medical interventions. Thus, we must be content that they will do so with some (usually high) degree of probability. Therefore, some degree of *uncertainty* always pervades clinical medical science, and the focus is on minimizing this to the extent possible. The generalizability of RCT results will always be a scientific inference based, at least in part, on the facts concerning the specific individuals and interventions under test and the relative comparability of a target population. However, due to the nature of the science involved, I would submit that the results of RCTs in clinical medical science are eminently more generalizable, and the scope to be not nearly so narrow, as Cartwright seems to suggest for RCTs in general. Indeed, I would argue, at least with respect to RCTs in clinical medical science, Cartwright's statement that "External validity for RCTs is hard to justify" (2007, 19) seems itself to be hard to justify.

7.7   Why the Hierarchical Pyramid of EBM Is Justified

I have argued that RCTs and the further studies that they spawn, systematic reviews of RCTs and meta-analyses of RCTs, deserve to be given the greatest weight in any system evaluating evidence from various types of studies in clinical medical science. The central issue in any such approach, I have argued, is *accuracy*. An accurate study is one in which the investigators actually measure what they believe they are measuring (validity), and are measuring it precisely. It is one that is free of bias and confounding. *The RCT is the most accurate method currently available in clinical medical science for achieving validity in experiments on groups of patients of sufficient size to warrant a controlled trial.* This is so because it is designed as an experiment under which the investigators are able to exert maximum control over any potential factors that might bias or confound the results. The known factors can be evenly distributed by design. Unknown factors can be given the highest probability of being evenly distributed by randomization.

Observational studies are more prone to biases and confounding because the variables under study are not under direct control of the investigators. Cohort studies are believed to be less error-prone than case-control studies, since in many instances cohorts can be observed over time and outcomes recorded as they occur, and often potentially biasing and confounding factors can be identified and attempts made to control for them in analyses. Bias from lack of information is ever present. In case-control studies, case and control subjects are usually identified after the disease or other health outcome of interest has occurred, and usually at least in part is based on medical records that are often biased due to errors or lack of information. Even when subjects fill out questionnaires or are personally interviewed, experience has shown that faulty memory is always a concern.

Cross-sectional studies have threats to validity that are thought to exceed other observational studies since information on exposure and disease status, in addition to other factors, is usually not available prior to conduct of the study. Thus methods for controlling such factors, such as matching and stratification, are also usually not available. Multivariable statistical methods are often used for hypothesis testing, and as the blood pressure study among Japanese males by Ueshima et al. (1984) described earlier demonstrates, important determinants of the outcome variable may not even be measured. And, as already noted, there are problems associated with determining whether exposure preceded disease, and problems associated with measuring prevalence rather than incidence. And, of course, all observational studies suffer from our inability to control for *unknown* factors.

Where would the evidence that I have labeled Level I and II in the studies that led to the RCTs in early breast cancer fit in the hierarchy? To determine this, one thing that needs to be considered is validity. How comparable are the cases in the reported series to their implicit or explicit controls? In the Level I study by Mustakallio (1954), there was no comparison group. No analysis was performed, only simple description. Any comparison group would have to be implicit. This is a case series by one investigator and would rate relatively low on any validity scale, since we do not know how a truly comparable control group would have fared. Level II evidence is a little better, since there is *some* attempt made to recognize possible biases. Nevertheless, the threats to validity are numerous since the treatment groups are rarely comparable, and so may perhaps rate somewhat better than Level I evidence.

Both the Level I and II studies are best classed, I believe, as *case series*, which, like case reports, are largely *descriptive*. Their value lies mostly in their ability to *generate* hypotheses. In this respect they are similar to descriptive statistics in epidemiology, where, for example, one might compare the incidence rates of breast cancer in Asia to the rates in the U.S., Canada, or Northern Europe, where in the latter three areas they are some six times higher (Henderson et al. 1996, 1023). Hypotheses can then be developed as to why the difference is so great, which can then be tested in further studies. Studies in which hypotheses are *tested* are usually referred to as *analytic*: RCTs, cohort studies, case-control studies, and many cross-sectional studies in clinical medical science are analytic.

The Level I and II studies were not analytic, and, like case reports, are at the bottom of the hierarchical pyramid. Certainly data of this poor quality are incapable of being convincing to the medical community at large, notwithstanding the opinions expressed by the authors of those studies. To convince clinicians in any great number to change from one therapy to another requires valid evidence that the new therapy is at least as effective, if not more effective, than what it is replacing. This comes from analytic studies, and this is the type of evidence that EBM values most highly and strives to obtain.

Adoption or acceptance of any system of evidence appraisal by study type must recognize that the relative weights assigned to the various types of study result from a generalization. It implies that *generally*, or *in the aggregate*, study types higher in the hierarchy are more valid than those lower. More precisely, they are *probably* more valid. It does not imply that *every* RCT, for example, is more valid than every study in a type lower in the hierarchy. Each study, RCT or not, must be evaluated on its own merits. As Greenhalgh notes, "... not even the most hard-line protagonist of EBM would place a sloppy meta-analysis or an RCT that was seriously methodologically flawed above a large, well-designed cohort study" (2010, 44). Given these considerations, the hierarchy or ranking by Greenhalgh (2010, 18; 43–44) seems well justified.

However, before leaving the subject of the EBM pyramid, it should be noted that the pyramid and suggested ranking of evidence presented earlier (Greenhalgh 2010, 18; 43–44) have been criticized and efforts have been made at improvement (e.g. Murad et al. 2016; Shaneyfelt 2016). In a review of recent progress in EBM, Djulbegovic and Guyatt (2017) argue that the GRADE system of ranking evidence (Guyatt et al. 2008) provides a much more sophisticated hierarchy of evidence and that GRADE is gaining in importance in the EBM community and among other users. No doubt effort at continued refinement in ranking evidence will be made as EBM further matures.

References

Adair, Frank E. 1943. The role of surgery and irradiation in cancer of the breast. *Journal of the American Medical Association* 121: 553–559.

Byar, David P., Richard M. Simon, William T. Friedewald, James J. Schlesselman, David L. DeMets, Jonas H. Ellenberg, Mitchell H. Gail, and James H. Ware. 1976. Randomized clinical trials. Perspectives on some recent ideas. *New England Journal of Medicine* 295: 74–80.

Caldwell, Glyn G., Delle B. Kelley, and Clark W. Heath Jr. 1980. Leukemia among participants in military maneuvers at a nuclear bomb test. A preliminary report. *Journal of the American Medical Association* 244: 1575–1578.

Cartwright, Nancy. 2007. Are RCTs the gold standard? *BioSocieties* 2: 11–20.

———. 2009. Evidence-based policy: What's to be done about relevance? For the 2008 Oberlin philosophy colloquium. *Philosophical Studies* 143: 127–136.

Cope, Oliver, Chiu-An Wang, Ann Chu, Chiu-Chen Wang, Milford Schulz, Benjamin Castleman, John Long, and William D. Sohier. 1976. Limited surgical excision as the basis of a comprehensive therapy for cancer of the breast. *American Journal of Surgery* 131: 400–407.

Cutler, Sidney J., and Fred Ederer. 1958. Maximum utilization of the life table method in analyzing survival. *Journal of Chronic Diseases* 8: 699–712.

Djulbegovic, Benjamin, and Gordon H. Guyatt. 2017. Progress in evidence-based medicine: A quarter century on. *Lancet* 390: 415–423.

Early Breast Cancer Trialists' Collaborative Group. 1995. Effects of radiotherapy and surgery in early breast cancer. An overview of the randomized trials. *New England Journal of Medicine* 333: 1444–1455.

Fisher, Ronald A. 1947. *The design of experiments*. 4th ed. Edinburgh: Oliver and Boyd.

Fisher, Bernard, Madeline Bauer, Richard Margolese, Roger Poisson, Yosef Pilch, Carol Redmond, Edwin Fisher, et al. 1985. Five-year results of a randomized clinical trial comparing total mastectomy and segmental mastectomy with or without radiation in the treatment of breast cancer. *New England Journal of Medicine* 312: 665–673.

Fisher, Bernard, Carol Redmond, Roger Poisson, Richard Margolese, Norman Wolmark, Lawrence Wickerham, Edwin Fisher, et al. 1989. Eight-year results of a randomized clinical trial comparing total mastectomy and lumpectomy with or without irradiation in the treatment of breast cancer. *New England Journal of Medicine* 320: 822–828.

Freedman, Benjamin. 1987. Equipoise and the ethics of clinical research. *New England Journal of Medicine* 317: 141–145.

Freireich, Emil J., and Edmund A. Gehan. 1979. The limitations of the randomized clinical trial. In *Methods in cancer research Vol. XVII*, ed. Vincent T. DeVita Jr. and Harris Busch, 277–310. New York: Academic.

Green, Stephanie, Jacqueline Benedetti, and John Crowley. 1997. *Clinical trials in oncology*. New York: Chapman and Hall.

Greenhalgh, Trisha. 2010. *How to read a paper: The basics of evidence-based medicine*. 4th ed. Chichester: Wiley.

Guyatt, Gordon H., David L. Sackett, John C. Sinclair, Robert Hayward, Deborah J. Cook, Richard J. Cook, and for the Evidence-Based Medicine Working Group. 1995. Users' guide to the medical literature IX. A method for grading health care recommendations. *Journal of the American Medical Association* 274: 1800–1804. [Published erratum appears in *Journal of the American Medical Association* (1996) 275: 1232.].

Guyatt, Gordon H., Andrew D. Oxman, Gunn E. Vist, Regina Kunz, Yngve Falck-Ytter, and Holger J. Schünemann. 2008. GRADE: What is "quality of evidence" and why is it important to clinicians? *British Medical Journal* 336: 995–998.

Haffty, Bruce G., Neal B. Goldberg, Marie Rose, Barbara Heil, Diana Fischer, Malcolm Beinfield, Charles McKhann, and Joseph B. Weissberg. 1989. Conservative surgery with radiation therapy in clinical stage I and II breast cancer. Results of a 20-year experience. *Archives of Surgery* 124: 1266–1270.

Henderson, Brian E., Malcolm C. Pike, Leslie Bernstein, and Ronald K. Ross. 1996. Breast cancer. In *Cancer epidemiology and prevention*, ed. David Schottenfeld and Joseph F. Fraumeni Jr., 2nd ed., 1022–1039. Oxford: Oxford University Press.

Howson, Colin, and Peter Urbach. 2006. *Scientific reasoning. The Bayesian approach*. 3rd ed. Chicago: Open Court.

Jacobson, Joan A., David N. Danforth, Kenneth H. Cowan, Teresa D'Angelo, Seth M. Steinberg, Lori Pierce, Marc E. Lippman, Allen S. Lichter, Eli Glatstein, and Paul Okunieff. 1995. Ten-year results of a comparison of conservation with mastectomy in the treatment of stage I and II breast cancer. *New England Journal of Medicine* 332: 907–911.

Kendall, Maurice, Alan Stuart, and J. Keith Ord. 1983. *The advanced theory of statistics*. Vol. 3. 4th ed. New York: Macmillan.

Ketcham, Alfred. 1986. Invited Commentary. *World Journal of Surgery* 10: 1019–1020.

Mantel, Nathan. 1966. Evaluation of survival data and two new rank order statistics arising in its consideration. *Cancer Chemotherapy Reports* 50: 163–170.

Mayo, Deborah G. 2005. Evidence as passing severe tests: Highly probable versus highly probed hypotheses. In *Scientific evidence. Philosophical theories and applications*, ed. Peter Achinstein, 95–127. Baltimore: Johns Hopkins University Press.

Murad, M. Hassan, Noor Asi, Mouaz Alsawas, and Fares Alahdab. 2016. New evidence pyramid. *Evidence Based Medicine* 21: 125–127.

Muss, Hyman B., Donald A. Berry, Constance T. Cirrincione, Maria Theodoulou, Ann M. Mauer, Alice B. Kornblith, Ann H. Partridge, et al. 2009. Adjuvant chemotherapy in older women with early-stage breast cancer. *New England Journal of Medicine* 360: 2055–2065.

Mustakallio, S. 1954. Treatment of breast cancer by tumour extirpation and roentgen therapy instead of radical operation. *Journal of the Faculty of Radiologists* 6: 23–26.

Papineau, David. 1994. The virtues of randomization. *British Journal for the Philosophy of Science* 45: 437–450.

Recht, Abram, James L. Connolly, Stuart J. Schnitt, Blake Cady, Susan Love, Robert T. Osteen, W. Bradford Patterson, et al. 1986. Conservative surgery and radiation therapy for early breast cancer: Results, controversies, and unsolved problems. *Seminars in Oncology* 13: 434–449.

Roush, Sherrilyn. 2009. Randomized controlled trials and the flow of information: Comment on Cartwright. *Philosophical Studies* 143: 137–145.

Sackett, David L., William M.C. Rosenberg, J.A. Muir Gray, R. Brian Haynes, and W. Scott Richardson. 1996. Evidence based medicine: What it is and what it isn't. *British Medical Journal* 312: 71–72.

Shaneyfelt, Terence. 2016. Pyramids are guides not rules: The evolution of the evidence pyramid. *Evidence Based Medicine* 21: 121–122.

Simon, Richard M. 2008. Design and analysis of clinical trials. In *Cancer: Principles and practice of oncology*, ed. Vincent T. DeVita Jr., Theodore S. Lawrence, and Steven A. Rosenberg, 8th ed., 578–589. Philadelphia: Lippincott Williams and Wilkins.

Spitalier, J.M., J. Gambarelli, H. Brandone, Y. Ayme, D. Hans, J.M. Brandone, C. Bressac, et al. 1986. Breast-conserving surgery with radiation therapy for operable mammary carcinoma: A 25-year experience. *World Journal of Surgery* 10: 1014–1019.

Straus, Sharon E., Paul Glasziou, W. Scott Richardson, and R. Brian Haynes. 2011. *Evidence-based medicine. How to practice and teach it*. 4th ed. Edinburgh: Churchill Livingstone.

Thomas, Patrick R.M., and Anne S. Lindblad. 1988. Adjuvant postoperative radiotherapy and chemotherapy in rectal carcinoma: A review of the Gastrointestinal Tumor Study Group experience. *Radiotherapy and Oncology* 13: 245–252.

Ueshima, Hirotsugu, Takashi Shimamoto, Minoru Iida, Masamitsu Konishi, Masato Tanigaki, Mitsunori Doi, Katsuhiko Tsujioka, et al. 1984. Alcohol intake and hypertension among urban and rural Japanese populations. *Journal of Chronic Diseases* 37: 585–592.

Veronesi, Umberto, Roberto Saccozzi, Marcella Del Vecchio, Alberto Banfi, Claudio Clemente, Mario De Lena, Giuseppe Gallus, et al. 1981. Comparing radical mastectomy with quadrantectomy, axillary dissection, and radiotherapy in patients with small cancers of the breast. *New England Journal of Medicine* 305: 6–11.

Williams, I.G., R.S. Murley, and M.P. Curwen. 1953. Carcinoma of the female breast: Conservative and radical surgery. *British Medical Journal* 2: 787–796.

Worrall, John. 2007. Evidence in medicine and evidence-based medicine. *Philosophy Compass* 2 (6): 981–1022.

Chapter 8
Ethics and Evidence: Is Evidence from Randomized Controlled Trials Necessary to Firmly Establish a New Therapy?

Abstract In this chapter, I explore ethical issues that may arise in conducting randomized controlled trials to test therapeutic hypotheses, and address the question of whether randomized controlled trials are always necessary to firmly establish a new therapy. I illustrate the ethical issues by discussing several studies of extracorporeal membrane oxygenation therapy in newborn infants, and subsequently describe two cases in which a new cancer therapy was firmly established without randomized trial data: combination chemotherapy and radiation therapy in cancer of the anal canal, and multi-drug chemotherapy for disseminated carcinoma of the testis. I conclude that considerable opinion exists that the primary duty of physicians (*qua* physicians) is to the patients under their care and that conducting research is secondary, notwithstanding the immense benefit of medical research to society. I also conclude that randomized controlled trial data are sometimes unnecessary when other convincing data are available.

8.1 General

It seems to have become generally accepted among the medical community that evidence from RCTs is the most convincing type of evidence for establishing a new therapy. For many, it would appear that the *only* reliable evidence for this purpose comes from RCTs. Thus Tukey, for example, states, "Many of us are convinced, by what seems to me to be very strong evidence, that the only source of reliable evidence about the usefulness of almost any sort of therapy or surgical intervention is that obtained from well-planned and carefully conducted randomized, and, where possible, double-blind clinical trials" (1977, 679). And, Cowan writes, "With some exceptions, participation of any group of patients in a nonrandomized trial is wholly unjustified and unethical since nothing can be learned from it" (1981, 10).

Greenhalgh has maintained that RCTs are unnecessary when a clearly successful intervention for an otherwise fatal condition is discovered (2010, 39). But what of the more usual case, where it is claimed that based on evidence that at least by some may be regarded as preliminary, a new therapy is equivalent to, or superior to, an

© Springer Nature Switzerland AG 2020

J. A. Pinkston, *Evidence and Hypothesis in Clinical Medical Science*, Synthese Library 426, https://doi.org/10.1007/978-3-030-44270-5_8

already established therapy for some condition? Must RCT evidence be acquired to establish the new therapy?

I will argue that the answer to the above question is, at least sometimes, "no," and illustrate this by discussing three examples: extracorporeal membrane oxygenation (ECMO) therapy for the treatment of respiratory failure in newborn infants, combination radiation therapy and chemotherapy for carcinoma of the anal canal, and multi-drug chemotherapy for disseminated testicular carcinoma. In the ECMO case, although RCTs were done, I will argue that they were unnecessary to establish ECMO as superior to currently available alternative therapies. This case raised serious ethical issues and has been previously discussed by others.[1] In the case of carcinoma of the anal canal, combined radiation therapy and chemotherapy became firmly established over surgery as initial treatment without an RCT being done. Similarly, three-drug chemotherapy became firmly established as standard therapy in disseminated testicular carcinoma without an RCT.

8.2   ECMO

8.2.1   Background and RCTs of ECMO

Respiratory failure is one of the major medical problems in newborn infants, and is a common cause of death in this age group. Hyaline membrane disease accounts for most of the cases; other causes include meconium aspiration syndrome, neonatal sepsis, and persistent fetal circulation syndrome (Bartlett et al. 1982).

Prior to the 1980s, conventional therapy consisted mainly of the use of a tracheal tube and mechanical ventilation, with supplemental oxygen. Most infants did well on this regimen, but a minority, between 5% and 10%, failed to respond and died of respiratory failure. Another 10% developed bronchopulmonary dysplasia, a disabling lung condition thought to be due to the pressure and oxygen used for treatment. During the 1970s and early 1980s, Bartlett et al. (1982) developed ECMO as an alternative therapy for respiratory failure in the newborn. The procedure involves the use of a modified heart-lung machine that can support gas exchange for days or even weeks until the neonatal lung has recovered. Under local anesthesia, the right atrium is cannulated through the right internal jugular vein and blood is passed extracorporeally through tubing connected to a source of oxygen, a membranous lung for gaseous exchange, a heat exchanger, a heparin infusion pump, and other supporting elements. The oxygenated blood is then passed back into the infant's aortic arch via a cannula leading from the right common carotid artery. This apparatus functionally bypasses the heart and lungs and allows the lungs to "rest," thus preventing bronchopulmonary dysplasia and saving the lives of some of these patients (Bartlett 1984).

---

[1]See, e.g., Royall (1991), Truog (1993), and Worrall (2008).

Bartlett et al. (1982) reported a series of 45 cases of neonatal respiratory distress that they had treated during the preceding eight years using ECMO. The patients had been referred by neonatology colleagues who identified them as unresponsive to maximal therapy with less than a 10% chance of survival. They were selected from approximately 1500 seriously ill infants, and all were receiving 100% oxygen with mechanical ventilation. Twenty-five of the 45 infants survived (56%). The investigators considered this experience as part of a phase 1 trial, and believed that a prospective controlled randomized "phase 2" trial was needed to better establish ECMO as superior to conventional therapy.

In 1985 Bartlett and colleagues reported the results of their prospective randomized study. Criteria were established to select patients with severe respiratory failure with a mortality risk of at least 80%. The study design did not employ the more usual RCT method where subjects are assigned randomly to experimental and control arms in approximately equal proportions, but rather to a "randomized play-the-winner" statistical method (Zelen 1969; Wei and Durham 1978). The procedure is equivalent to the following: The treatments are coded A and B and two balls are placed in an urn, one labeled A, and the other B. For each patient, the assigned treatment is determined by which ball is drawn from the urn. After the first ball is drawn and the indicated treatment is administered, the ball is returned to the urn and a new ball is added. If the treatment was successful, the new ball carries the same letter; if not, it carries the other letter. When ten balls of one type have been added, that treatment is considered the winner. Thus a study of this type requires at least ten and at most 19 subjects (Royall 1991, 59).

The design provided that under the assumption that one treatment was substantially better than the other, the probability is very high that the randomized play-the-winner rule will select as the winner the treatment that is actually better. For the *a priori* probability that $P_A \geq 0.8$ and $P_A - P_B > .04$, where P_A denotes the probability of survival when the infant receives the better treatment and P_B the corresponding probability when the infant receives the poorer treatment, the probability of selection of the best treatment is at least 0.95. For the probabilities actually thought by the investigators to hold, namely $P_A = 0.9$ for ECMO and $P_B = 0.1$ for conventional therapy, the probability of selecting the better treatment is even greater (Bartlett et al. 1985, 484–485).

Twelve patients entered the study. The first patient was randomly assigned to ECMO and survived. The second patient was randomly assigned to conventional therapy and died. The next ten patients were assigned to ECMO and survived. At study termination, there was one control patient who had died, and 11 ECMO patients, all of which survived. They also included in their report that, since study termination, ten additional patients that met entry criteria for their study were seen at their institution. Eight were treated with ECMO and all survived. Two infants were not treated with ECMO, and both died. The authors concluded that based on 19 consecutive successes with ECMO, the lower 99% one-sided confidence interval on the survival with ECMO is 78.5. They state, for a 1% significance level, that "... the null hypothesis that the survival probability is the same for ECMO as for conventional therapy would be rejected in favor of a higher survival probability for ECMO for any specification of a survival probability for conventional therapy less than 78.5" (Bartlett et al. 1985, 485). Since this survival probability is well above that observed in the past for this population when given conventional therapy, the investigators concluded that ECMO was statistically superior.

Nevertheless, the investigators were aware that their study design was unconventional, and they note that although the randomized play-the-winner statistical technique had been introduced in 1969, it had not previously been used in a clinical study (1985, 480). After all, only one patient had been randomized to the control arm. Perhaps anticipating criticism, they state, "In retrospect, it would have been better to begin with two or three pairs of balls, which would have resulted in more than one control patient" (1985, 484).

The study was criticized by Ware and Epstein, who argued that "... the results are not completely convincing. Why not? Because only one patient received the standard therapy, so that the interpretation of the study depends strongly on the belief that eligible patients would have experienced poor survival in the absence of [ECMO]" (Ware and Epstein 1985, 850). They conclude: "Further randomized clinical trials using concurrent controls and addressing the ethical aspects of consent, randomization, and optimal care will be difficult but remain necessary"(Ware and Epstein 1985, 851).

O'Rourke et al. (1989) reported a later RCT. Thirty-nine infants were enrolled, and the study was designed so that a maximum of four deaths were allowed in either the conventional therapy or ECMO group. The first 19 patients were randomly assigned to conventional therapy or ECMO. Nine patients received ECMO, and all survived. Ten patients received conventional therapy. Of these, six survived and four died. The RCT portion of the study was terminated at this point (phase I). The next 20 patients were assigned to ECMO (phase II). Of these, 19 survived and one died. The study was then terminated. Four deaths had occurred among ten infants given conventional therapy, and one death had occurred among the 29 patients given ECMO. Statistical analysis of the data by the authors showed that the results represented ECMO as the superior therapy (p < .05), which they argued was a conservative estimate of efficacy.

Pocock (1993) drew attention to the paucity of data when the RCT (phase I) was stopped, with only 19 patients having been randomized. Many believed that a larger RCT was needed, and subsequently a collaborative randomized trial was undertaken in the United Kingdom (UK Collaborative ECMO Trial Group 1996). Between 1993 and 1995, 185 neonates with severe respiratory failure were enrolled from 55 hospitals. Those randomized to ECMO were referred to one of five centers with ECMO facilities. Those randomized to conventional therapy continued to receive such therapy at their original hospitals. Ninety-three of the 185 patients were randomized to ECMO, and 92 were randomized to conventional therapy. Recruitment to the trial was stopped early on the advice of the independent data-monitoring committee, since the data showed a clear advantage with ECMO. Thirty of the 93 neonates that received ECMO died (32%) and 54 of the 92 infants randomized to conventional therapy died (59%). The relative risk was 0.55 (95% CI 0.39–0.77; p = 0.0005), which is equivalent to one extra survivor for every three to four infants allocated to ECMO.

8.2.2 Were the ECMO RCTs Necessary?

The ECMO RCTs were, I will argue, unnecessary to establish ECMO as the preferred therapy for the class of neonates with respiratory failure that were considered eligible for randomization. The reasons involve both ethical and epistemic considerations.

Ethical considerations are an important element in the design and conduct of RCTs in clinical medical science because RCTs are experiments and the subjects are human beings. Any new therapy being advanced involves the belief that it is superior, or at least not inferior, to existing therapies before it can ethically be tested, regardless of study type. This is true whether the proposed new therapy is being studied, for example, in a small group of patients to establish toxicity profiles, or in a larger nonrandomized study of the efficacy of the new therapy in comparison with historical controls. Reasonable evidence of safety and efficacy must exist to advance a new therapy to the stage of an RCT. And as I have argued, and the EBM movement also maintains, RCTs provide the best evidence, i.e., the strongest epistemic underpinning, for confirming therapeutic hypotheses.

The emphasis on the acquisition of sound scientific evidence to undergird clinical decision-making, the main thrust of the EBM movement, is relatively recent. As noted previously, Worrall (2007, 986) has observed that most current medical therapies have not been established with RCT evidence. Goodman (2003, 6) notes that only about 10–25% of health care is based on high-quality or gold-standard evidence, and it is estimated that only about 50% of current medical practice is evidence-based (McGlynn et al. 2003, 2643). Thus, the basis for most medical therapy comes from the experience of clinicians, supported by research of various types short of an RCT.

Clinicians are ethically charged with administering, or at least recommending, what they believe to be the best available therapy for each individual patient. It must also be recognized that clinicians have many other ethical obligations, among them to family and society in general, in addition to those to the individual patient. But in the ordering of ethical obligations, the clinician, *qua* clinician, is widely regarded as being obligated to place primary consideration on the health of the individual patient under his or her care. This is promulgated in various codes: for example, in the 1948 Declaration of Geneva of the World Medical Association, it is affirmed that, "the health of my patient will be my first consideration" (Beauchamp and Childress 1994, 441). Schafer (1982, 720) notes that, "In his traditional role as healer, the physician's commitment is exclusively to his patient." Fried (1974, 50) states that, "The traditional concept of the physician's relation to his patient is one of unqualified fidelity to that patient's health." He calls this the *personal care concept.* And, Pelligrino (1979, 114) states, "Surely the first order of responsibility for clinicians must remain with the patients they undertake to treat. Here, the moral imperatives are clear: competence of the highest order, integrity, compassion."

But, arguably, physicians also have an obligation to work to improve the quality of the care and treatment that patients on the whole receive. Physician-investigators thus generate and participate in various levels of research, and physicians in general are encouraged to seek to enroll eligible patients in ongoing studies. Does this mean, however, that physicians or physician-investigators should participate in studies, including RCTs, in which it is possible, by study design, that at least some patients will receive a therapy that they believe is inferior? Or, might they be ethically obligated *not* to participate?

Under the ethical principle that the *primary* duty of the physician (*qua* physician) is to the health and well-being of his or her individual patient, it would seem that the Bartlett et al. (1985) and O'Rourke et al. (1989) RCTs were in breach of that principle, since it seemed that the investigators clearly believed that ECMO was superior to conventional treatment. That they so believed is suggested by at least two observations: the consent process used, and the trial designs that were selected.

In order to proceed with the trials, the investigators were obligated to obtain informed consent from the parents of the neonates. A frank and honest discussion of the risks and benefits of ECMO and conventional therapy would have included the results to date observed with those alternatives, particularly what results they (the parents) could expect under each alternative. The consent process employed for both RCTs used a randomization method advanced by Zelen (1979), in which eligible patients are randomized before consent is sought. Neonates randomized to conventional therapy would receive the same treatment anyway, it is reasoned, and thus they are arguably not part of the experiment. Thus, consent from these parents need not be obtained. Consent is only sought for the "experimental" therapy, in this case, ECMO. Bartlett et al. (1985, 484) contended that "…if consent is sought before randomization, the distraught family is presented with confusing treatment options which they cannot fully understand…"[2] O'Rourke et al. (1989, 959) stated that, "This [Zelen (1979)] method was chosen in the belief that discussing the possibility of ECMO therapy with families whose child did not ultimately receive ECMO would not benefit those families and would create additional emotional distress."

Another indication of ethical conflict was the unconventional trial designs that were used in the two studies. This was clearly done to minimize the number of infants assigned to conventional therapy. Bartlett et al. (1985) justified their "play-the-winner" design because of their anticipation that "…most ECMO patients would survive and most control patients would die, so significance could be reached with a modest number of patients," and because "It was a reasonable approach to the scientific/ethical dilemma … we were compelled to conduct a prospective, randomized study, but reluctant to withhold a lifesaving treatment from alternate patients simply to meet conventional random assignment technique" (1985, 480). O'Rourke

---

[2]Recall that the investigators believed (or at least assumed) that the survival probabilities were 90% with ECMO and 10% for conventional therapy (Bartlett et al. 1985, 485).

et al. (1989, 962) rejected a fixed sample size design "...because of the potential for a large difference in mortality rates...".

Thus the investigators were seemingly not in a state of "personal equipoise." They apparently believed that ECMO was superior. But did clinical equipoise exist? Clinical equipoise, the existence of uncertainty and disagreement among the expert medical community about which treatment is better, is, as I argued earlier in relation to the mastectomy versus irradiation trials in early breast cancer, what ethically justifies the RCT. When the ECMO trials were initiated, was the relevant expert medical community in clinical equipoise? In other words, was there sufficient evidence already of ECMO's superiority?

When the trials were begun, it seems difficult to imagine that clinicians involved in the care and treatment of moribund infants in respiratory distress that were on conventional therapy and not responding, and thus with an expected high mortality rate, the very class of infants that were eligible for randomization, did not view ECMO as potentially lifesaving and clearly superior to continuing on conventional treatment. Indeed, they were referring such cases for ECMO when it was feasible to do so.

Most RCTs do not raise ethical concerns among the research subjects or clinicians involved. Most reservations are relatively minor and can be satisfactorily resolved at the institutional review board (IRB) level of oversight. As Truog (1993, 524) notes, "Few criticize the RCT that seeks to identify the best antibiotic for treating acute otitis media or the best antacid for peptic ulcer disease."

The ethical problems surrounding the ECMO RCTs derive from the fact that the therapies under consideration are potentially lifesaving. The neonates selected for the ECMO trials were judged to have a high mortality risk; for the Bartlett et al. (1985) study it was at least 80%, and was estimated to be about 75% for the O'Rourke et al. (1989) study. Data on infants treated with ECMO showed about a 75% survival rate (Bartlett 1984, 140) when the Bartlett et al. (1985) study was initiated, and an ECMO Registry report of 715 infants treated with ECMO showed an overall 81% survival rate when the O'Rourke et al. (1989) study was undertaken (Toomasian et al. 1988, 141).

How much evidence of efficacy of a potentially lifesaving therapy must exist for that therapy to become the preferred therapy? Is RCT evidence required? As Royall (1991, 60) notes, "Science desires randomized clinical trials, it does not demand them." And Fried states unequivocally that "...the claims for the RCT have been greatly, indeed preposterously overstated. The truth of the matter is that the RCT is one of many ways of generating information, of validating hypotheses. The proponents of the RCT, however, have elevated what is in theory a frequent (though by no means universal) advantage of degree into a gulf as sharp as that between the kosher and the non-kosher" (1974, 158). The investigators were themselves apparently convinced of the superiority of ECMO. And it appears that the parents of infants that were offered a choice between ECMO and conventional therapy were equally convinced. In the O'Rourke et al. (1989) study, after randomization, all

29 patients' parents who were approached for ECMO gave their consent for ECMO.[3]

The O'Rourke et al. (1989) study led to a debate over the unusual statistical design and the ethical questions it raised. This resulted in a rare reprimand of the Boston Children's Hospital's IRB by the National Institutes of Health (NIH) for failing to ensure that all subjects in a clinical trial were informed (Marwick 1990, 2420).

The RCT carried out in the UK was done because it was believed that ECMO was controversial in view of the varying interpretations of the available evidence. The UK trial organizers viewed the studies by Bartlett et al. (1985) and O'Rourke et al. (1989) as inconclusive. Most of the claims about ECMO were based on case series and other studies with historical controls, which, although suggesting large reductions in mortality, were carried out at a time when neonatal death rates were falling. Neonatal ECMO was introduced into the UK in 1989, but some clinicians were reluctant to refer potential neonates for ECMO because of concerns that any improved survival from the technique might be offset by an increase in long-term disability. Others were concerned about the costs of ECMO, which exceeded those of conventional therapy, while questions about its clinical effectiveness and cost-effectiveness persisted. Based on these factors, British clinicians agreed to limit neonatal ECMO to use within an RCT.

Consent was obtained from the parents of neonates in the usual manner in the UK trial, before randomization (Field 1995, 1370). Also, a conventional study design was used wherein patients were randomized in approximately equal numbers to conventional therapy or ECMO. The trial was carried out between 1993 and 1995, well after the results of the Bartlett et al. (1985) and O'Rourke et al. (1989) RCT results were available. Subsequently, the trial was criticized as unnecessary and unethical.

For example, Lantos (1997, 265) pointed out that by 1993, when the UK trial was initiated, more than 7500 neonates had been treated with ECMO in 75 programs in the United States and 17 programs in other countries. He states, "More certainty is always better than less certainty, but at some point we need to decide that we are certain enough" (1997, 266). Would another trial of ECMO to confirm the results of the UK trial be ethical? He says, "I think that the data that were available in the early 1990s on the benefits of ECMO were convincing... If I was on an ethics review panel, I would not have approved the trial" (1997, 267–268). In the same vein, he also suggests that since the "default" choice for parents with infants eligible for enrollment in the trial was conventional therapy, and the only chance to receive ECMO was to enroll in the trial, there was an implicit element of coercion in the trial design.

Other ethical issues arise when either the conventional or experimental therapy (or both) is rapidly evolving (Truog 1993, 525–526). In addition to being a

---

[3]This would seem to assume that these parents were fully informed of the risks and benefits of the alternative treatments.

potentially lifesaving therapy, during the period of the trials, ECMO was rapidly developing. For example, many institutions were switching from veno-arterial to a veno-venous technique utilizing the jugular vein for both withdrawing blood and returning it to the body. ECMO apparatus not requiring anticoagulants was under development. And, perhaps in response to avoiding the cost and complexity of ECMO, improvements in conventional therapy were further reducing the mortality rate associated with those therapies. Wung et al. (1985) reported treating 15 seriously ill neonates in respiratory failure with modifications in ventilatory therapy focused on reducing barotrauma. ECMO was not used, and all survived. Schapira and Solimano (1988) reported two deaths among 13 neonates (a mortality rate of 15%) with severe respiratory distress due to meconium aspiration syndrome treated between 1983 and 1987 that met criteria for ECMO, but were treated with conventional therapy. A retrospective review by Dworetz et al. (1989) of severely ill neonates that met ECMO trial entry criteria but were treated with conventional therapy showed an improvement in survival of those treated between 1980–1981 and those treated between 1986–1988. One of six patients survived in the earlier period (17%), whereas nine of ten patients (90%) survived in the later period, possibly due to changes in ventilatory therapy. Granted that the numbers from these case series were small and the results not generalizable, they do indicate that efforts at progress were being made. And, in the O'Rourke et al. (1989) study, six of the ten infants randomized to conventional therapy survived (60%), which was higher than expected.

RCTs of rapidly evolving therapies pose at least two ethical problems, one related to the requirements of RCTs, the other related to relevance (Truog 1993, 525–526). RCTs are usually designed to keep the treatments constant, and may take years to complete. Thus, innovations that occur while the study is underway may not be available to study subjects. Patients on either the control or experimental arm (or both) may wind up receiving inferior therapy compared to similar patients being treated outside the study. Also, the trial itself may retard the development of new approaches and technologies, particularly among the institutions involved in the study, as study results are awaited.

The other problem is relevance. Most consider it to be a fundamental ethical requirement that RCTs have the potential to generate useful knowledge. Cowan (1981, 10) for example, states that, "...good research design requires that any proposed clinical trial be scientifically sound and capable of yielding generalizable data; a study lacking these characteristics is inherently unethical." Over the time period of the studies, the mortality rate of conventional therapy changed markedly, from 80% to perhaps as little as 10%, as noted above. This severely questions the leading assumption of the trials: a high ($\geq$80%) mortality rate in neonates treated with conventional therapy. Thus, the information generated from such trials may be obsolete and not useful when the results become available.

If RCTs are not practical to generate evidence on which to base clinical decisions for these rapidly evolving therapies, what is the best method to accumulate and evaluate the evidence that is being acquired? One suggestion with seeming merit that has been advanced is establishment of a prospective observational database (Truog

1993, 526; Berry 1989, 309–310). For neonates with severe respiratory distress, for example, clinicians would treat patients with the methods that they believe to be the most efficacious. No restrictions on how patients are treated are imposed. All participating institutions would send pertinent patient and treatment information to a central registry. Data on the effectiveness of various interventions would be periodically analyzed. Algorithms would be devised to assess outcomes on patients matched for prognostic factors.

Truog (1993) believes that such a registry would have obviated the need for the O'Rourke et al. (1989) RCT. The ECMO registry of 715 neonates published in 1988 by Toomasian et al. demonstrated an 81% survival with ECMO and indicated that ECMO was statistically superior to any other therapy with a survival rate of less than 78%. Had the registry included similar conventionally treated neonates, it would have shown the superiority of ECMO. Possibly, the UK trial could also have been avoided.

In view of the foregoing, it is arguable that the ECMO RCTs were unnecessary, since ECMO was clearly a potentially life-saving therapy. The EBM movement also considers RCTs for such life-saving treatments to be unnecessary. As previously noted, Greenhalgh (2010, 39) has so stated, and Sackett et al. (1996, 72) maintain that "...some questions about therapy do not require randomised trials (successful interventions for otherwise fatal conditions) or cannot wait for the trials to be conducted." Nowhere is the duty of the clinician to his or her individual patient stronger than when the patient's very life is at stake. Here, ethical considerations preclude the acquisition of evidence from the admittedly more epistemically desirable RCT. As the Physician's Oath of the World Medical Association states: "Concern for the interests of the subject [of research] must always prevail over the interests of science and society" (Beauchamp and Childress 1994, 441). And similarly, A.B. Hill (1963, 1047) says, "...the ethical obligation always and entirely outweighs the experimental."

8.3 Carcinoma of the Anal Canal

The anal canal is the terminal portion of the digestive tract and ends in association with sphincter musculature that control evacuation of the products of digestion. The columnar mucosa of the rectum transitions into a squamous histology, and squamous cell carcinomas include so-called basaloid, cloacogenic, and epidermoid carcinomas. These carcinomas are histologically distinct from the more frequently occurring rectal adenocarcinomas (Welton and Raju 2011, 344).

Until the 1970s the preferred treatment of squamous cell carcinoma of the anal canal was primarily surgical. Small tumors could usually be excised successfully without much morbidity, but larger tumors, which often invaded the sphincter musculature, required the more extensive and morbid abdominal perineal resection. This operation involved an intra-abdominal component and a perineal component, and resulted in removal of the distal rectum and anus, with closure of the perineal defect and a permanent colostomy. Local recurrence rates ranged from 27% to 47%, and five-year survival rates ranged from 40% to 70% (Welton and Raju 2011, 345).

In 1974 Nigro et al. reported the use of combined radiation therapy and chemotherapy in the form of 5-fluorouracil (5-FU) and mitomycin-C in the treatment of anal canal cancer. Three patients were administered the combined therapy as part of a planned preoperative program to be followed by abdominal perineal resection. The purpose of the preoperative regimen was to improve the local control and cure rates. Two patients completed the planned radiation therapy and chemotherapy, and no evidence of cancer was found in the surgical specimen obtained after their abdominal perineal resections. The third patient completed the radiation therapy and chemotherapy, but refused surgery. There was no evidence of cancer in the patient 14 months after treatment. The authors included in their report a woman with metastases to the liver from cloacogenic carcinoma from the anal canal treated with the same chemotherapy regimen but with a lower radiation dose to the liver. Within a few weeks the enlarged liver had shrunk to less than normal size, and there was no evidence of residual disease either by biopsy or laparoscopy. The authors did not claim this to be curative, but noted that they had not seen such a dramatic response to any therapy for this condition before.

By 1976 reports of the use of the "Nigro regimen" (as it later came to be called) began to appear (e.g., Newman and Quan 1976), although, except for the less commonly appearing smaller tumors, abdominal perineal resection alone was still advocated as definitive therapy (Wilson et al. 1976; Golden and Horsley 1976). Newman and Quan (1976), for example, reported three patients with surgically incurable epidermoid carcinoma of the anus treated with the Nigro regimen; one died during the course of the therapy, but the two others achieved apparent complete resolution of their local tumor: one patient was alive and well nearly one and a half years after initiation of therapy, and the other underwent abdominal perineal resection after completing the radiation and chemotherapy, with no residual carcinoma seen on pathological examination of the surgical specimen. The authors included a fourth patient in their report who had biopsy-proven pulmonary metastases treated with 5-FU and mitomycin-C. Six weeks later, a chest radiograph showed essentially complete disappearance of the metastatic nodules. The authors concluded that their experience suggested that multimodality therapy might increase salvage in even locally far-advanced and metastatic epidermoid anal carcinoma.

During this period, improvements in radiation therapy equipment and technique were occurring, and some studies were reporting more favorable outcomes for some patients when radiation therapy was added to surgery (e.g., Green et al. 1980). And, by 1980, further reports of the use of combined modality chemotherapy and radiation therapy were appearing that seemed to indicate a potentially major advance in the treatment of anal carcinoma. Sischy et al. (1980) reported ten patients with anal carcinoma confined to the anorectal area. Four patients received preoperative radiation therapy and chemotherapy consisting of mitomycin-C and 5-FU and subsequently underwent abdominal perineal resection. None had residual tumor on pathological examination of the surgical specimen. The other six patients were treated definitively with the chemotherapy and irradiation alone, without surgery.

These patients were also found to be free of disease, proved by biopsy. They found it impossible to predict the outcome of the treatment by the size of the original lesion, which was remarkable since tumor size is usually an important predictive factor for tumor response to therapy. They state: "... in instances of squamous cell carcinoma of the anus, if the lesion has disappeared completely at the end of treatment, adequate biopsies may be taken, and only in those instances in which there is residual tumor should abdominoperineal resection be performed. In this way, it is possible that a large number of patients with squamous cell carcinomas could be spared abdominoperineal resection" (Sischy et al. 1980, 370).

Also, in 1980 Cummings et al. reported six patients referred to the Princess Margaret Hospital in Toronto with anal canal cancer treated between May 1978 and August 1979 with radical radiation therapy plus 5-FU and mitomycin-C chemotherapy. No patients had surgery, and all had complete disappearance of their tumor within two months of completion of therapy. None showed any evidence of late recurrence, and they all retained anal continence with good control of bowel function by the anal sphincter musculature.

The remainder of the 1980s saw further reports appear from several centers using the new approach of preoperative 5-FU and mitomycin-C combined with radiation therapy, in which surgery was increasingly being reserved for patients that failed the preoperative regimen. It became customary to closely monitor patients after the preoperative regimen for any sign of recurrence, with biopsies done as needed. Improvements in radiation targeting and delivery were also occurring, and by 1993 Cummings editorialized that,

> The need for a randomized trial in which radical surgery would be compared with radiation therapy or radiation combined with chemotherapy, desirable though it may have been a decade ago, has now passed, and there can be little doubt that radiation-based protocols are at least as effective as surgery in terms of overall survival rates, and enable anorectal function to be preserved... (1993, 173).

The principle of using a combination of chemotherapy and radiation therapy as planned definitive therapy, with surgery reserved for the salvage of the minority of patients that failed this strategy, had essentially by this time become firmly established as preferred therapy without an RCT to test this hypothesis. Further questions arose, however, such as whether 5-FU or mitomycin-C, or both, could be omitted without compromising outcomes, or whether some other chemotherapy drug, such as cisplatin, could be substituted for the more toxic mitomycin-C. These questions *were* addressed with RCTs.

For example, to study whether the addition of 5-FU and mitomycin-C chemotherapy to irradiation was necessary, the European Organization for Research and Treatment of Cancer Radiotherapy and Gastrointestinal Cooperative Groups carried out an RCT to compare these approaches. One hundred ten patients from participating cancer centers in Israel and seven European countries were randomized to either radiation therapy alone or radiation therapy plus 5-FU and mitomycin-C chemotherapy. Results showed a significant increase in the complete remission rate from 54%

for radiation therapy alone to 80% for radiotherapy combined with 5-FU and mitomycin-C, leading to a significant improvement in locoregional control and colostomy-free survival (p = .02). The overall survival rate remained similar in both groups, due to the ability of surgery to salvage treatment failures (Bartelink et al. 1997).

Mitomycin-C is considered a relatively toxic chemotherapy drug. In addition to causing myelosuppression (lowering of blood counts), it is also known to have pulmonary, cardiac, hepatic, and renal toxicities, the latter of which can be life-threatening. To test the hypothesis that mitomycin-C could be omitted from the chemotherapy regimen, an RCT was performed in the U.S. Institutions in the Radiation Therapy Oncology Group and the Eastern Cooperative Oncology Group participated. Between 1988 and 1991, 310 patients were randomized to receive radiotherapy and 5-FU, or radiotherapy, 5-FU, and mitomycin-C. At four years, colostomy rates were lower (p = .002), colostomy-free survival was higher (p = .014), and disease-free survival was higher (p = .0003) in the group that received mitomycin-C. Toxicity was greater in the mitomycin-C group. The authors concluded that notwithstanding the increased toxicity, the use of mitomycin-C is justified, particularly in patients with large tumors (Flam et al. 1996).

To test the hypothesis that cisplatin could replace mitomycin-C, a large randomized trial with 649 evaluable patients was carried out in the U.S. in which several trial groups participated. The randomization was between radiotherapy, 5-FU, and mitomycin-C versus radiotherapy, 5-FU, and cisplatin. Five-year disease-free survival and five-year overall survival favored the group receiving mitomycin-C (p = .006 and p = .026, respectively). There was a trend toward statistical significance for colostomy-free survival (p = .05), the rate of locoregional failure (p = .087), and colostomy failure (p = .074). The authors concluded that the combination of 5-FU and mitomycin-C yielded a statistically significant, clinically meaningful improvement in disease-free survival and overall survival, and has borderline significance for colostomy-free survival, colostomy failure, and locoregional failure when compared to 5-FU and cisplatin. They also concluded that radiotherapy with 5-FU and mitomycin-C remains the preferred treatment for anal canal cancer (Gunderson et al. 2012).

When the results of the early studies showing the promise of radiation therapy combined with 5-FU and mitomycin-C chemotherapy in the treatment of anal cancer were made available, such as those by Nigro et al. (1974), Newman and Quan (1976), and others, why were one or more RCTs not performed to directly test the new approach against the established conventional therapy of abdominal perineal resection? Any answer would necessarily be speculative, but one plausible explanation lies in the dramatic, unexpected response of this tumor to the new approach, which was completely different from responses seen in the anatomically nearby rectal cancers or in other cancers of epithelial origin, such as lung or breast cancer. And, surgery for the salvage of failures of the new approach was still available.

8.4   Disseminated Carcinoma of the Testis

Carcinomas of the testis are mostly of germ cell origin, which are cells that are destined to become sperm cells. Histologically, carcinomas of the testis are mostly embryonal cell carcinomas, teratocarcinomas, choriocarcinomas, or some combination of these cell types. They are a disease of younger men, and are the most common solid tumors in men aged 20–34 years of age. The usual presentation is a nodule or painless swelling of the testicle. Initial treatment (and diagnosis) is accomplished by surgical removal of the affected testis. Unfortunately, in from 60% to 70% of cases, the disease is disseminated when the diagnosis is made. Before the advent of anti-cancer chemotherapy drugs, treatment of disseminated testicular carcinoma was largely unsuccessful and most patients died of their disease (Richie 1998).

Carcinoma of the testis was found early on to be moderately sensitive to some chemotherapy drugs. For example, in 1967 Wyatt and McAninch reported ten men with disseminated testicular carcinoma treated with methotrexate. Four achieved a complete remission, but none of the other six responded and all six died. In 1975, Samuels et al. reported 23 patients with disseminated testicular carcinoma treated with vinblastine and bleomycin. Nine of the 23 patients achieved a complete remission (39%) and eight achieved a partial remission (35%).

In 1977, Einhorn and Donohue reported 50 patients with disseminated testicular carcinoma treated with the three-drug combination of cisplatin, vinblastine, and bleomycin. Two patients died within one week of the initiation of chemotherapy, and a third patient died two weeks after the start of chemotherapy, all presumably due to massive tumor. All three of these patients had significant respiratory symptoms due to massive pulmonary metastases. This left 47 evaluable patients, and this regimen produced a complete remission in 35 patients (74%) and a partial remission in the other 12 (26%). Five of the patients with a partial remission became disease-free after surgical removal of residual disease, yielding an overall 85% disease-free status.

The three-drug regimen reported by Einhorn and Donohue (1977) became the "standard" or conventional treatment for disseminated testicular carcinoma. Further studies would be RCTs to test modifications of the regimen, for example to reduce toxicity or to improve remission and survival rates. In 1981, Einhorn et al. reported that five years after the Einhorn and Donohue (1977) study, 27 of their original 47 patients (57%) remained alive and disease-free, with a 19% relapse rate. Since the great majority of such patients relapse within three years of completing therapy, these 27 patients are presumed cured. In their report they presented the results of an RCT testing whether the addition of doxorubicin to the three-drug regimen improved results compared with the three-drug regimen alone, and whether "maintenance therapy," which is the continuation of some chemotherapy beyond the induction of remission, was of value.

A total of 184 consecutive patients were randomized to the three-drug regimen or the three-drug regimen plus doxorubicin. Those patients that achieved a complete remission or disease-free status following resection of residual disease that showed no viable tumor were further randomized to no maintenance therapy or maintenance therapy of monthly vinblastine for two years. Results showed no statistically significant differences in the groups. Thus, the original three-drug regimen remained the standard treatment. The results with the three-drug regimen were replicated in numerous institutions in the U.S., Canada, and Europe, as well as cooperative groups (Einhorn et al. 1981, 729).

Vinblastine produces significant neuromuscular toxicity, and etoposide had shown activity against testicular carcinoma in patients that had failed the three-drug regimen. In 1987, Williams et al. reported the results of an RCT comparing cisplatin and bleomycin plus either vinblastine or etoposide in disseminated testicular tumors. Among 244 patients that were evaluable for a response, 74% of those receiving the regimen including vinblastine and 83% of those receiving the regimen including etoposide became disease-free with or without surgery; however, this difference was not statistically significant. Survival among the etoposide group was higher (p = .048). In addition, the etoposide regimen showed statistically significant less toxicity. The regimen of cisplatin, etoposide, and bleomycin became the new standard therapy.

8.5 Conclusions

Perhaps the most important common characteristic in the ECMO, anal canal carcinoma, and disseminated testicular carcinoma examples is the dramatic improvement in outcomes provided by the new therapy compared to what was available before. In these cases, an RCT to establish this was arguably unnecessary. Glasziou et al. (2007) provide some other examples where dramatic effects have established some approaches without an RCT. And, as Miller and Joffe (2011, 479) point out, evidence of large effect sizes on the basis of early clinical studies is one criterion for approval of new oncology drugs by the U.S. Food and Drug Administration. For example, cisplatin was approved in 1978 for the treatment of testicular cancer, following the 1977 report by Einhorn and Donohue that established cisplatin, vinblastine, and bleomycin as the new standard therapy in disseminated testicular carcinoma. Indeed, in a review of oncology drug approvals between 1973 and 2006, of a total of 68 drugs that were approved, 31 were done so on the basis of studies that were nonrandomized (Tsimberidou et al. 2009, 6243).

How should such evidence be treated in the weight of evidence account? I have argued that when evaluating a new therapy, the RCT (and by extension, systematic reviews and meta-analyses of RCTs) provides, in general, the most epistemically desirable testing method *among study types*. It must be remembered that the ethical justification for the RCT is clinical equipoise, disagreement among the expert medical community as to the best treatment, and that the objective is social, to change standards of practice (Freedman 1987). Useful, generalizable, abstract knowledge is expected to result. In the ECMO, carcinoma of the anus, and disseminated testicular carcinoma instances, *compelling and convincing* evidence emerged almost serendipitously that the new approaches were superior to what was already available. It is worth emphasizing that what made the evidence so compelling was that the *majority* of patients fared better, and dramatically so, and thus clinical equipoise arguably did not exist.

In the weight of evidence account, the evidence from the ECMO and anal and testicular carcinoma examples must be considered strong. Thus, when responses occur that are clearly definitive, treatment approaches that have not been studied in an RCT can become the new standard approach and be incorporated into treatment guidelines by major organizations such as the NCCN and the American Society for Radiation Oncology.

References

Bartelink, H., F. Roelofsen, F. Eschwege, P. Rougier, J.F. Bosset, D. Gonzalez Gonzalez, D. Peiffert, M. van Glabbeke, and M. Pierart. 1997. Concomitant radiotherapy and chemotherapy is superior to radiotherapy alone in the treatment of locally advanced anal cancer: Results of a phase III randomized trial of the European Organization for Research and Treatment of Cancer Radiotherapy and Gastrointestinal Cooperative Groups. *Journal of Clinical Oncology* 15: 2040–2049.

Bartlett, Robert H. 1984, April. Extracorporeal oxygenation in neonates. *Hospital Practice* 19: 139–151.

Bartlett, Robert H., Alice F. Andrews, John M. Toomasian, Nick J. Haidue, and Alan B. Gazzaniga. 1982. Extracorporeal membrane oxygenation for newborn respiratory failure: Forty-five cases. *Surgery* 92: 425–433.

Bartlett, Robert H., Dietrich W. Roloff, Richard G. Cornell, Alice French Andrews, Peter W. Dillon, and Joseph B. Zwischenberger. 1985. Extracorporeal circulation in neonatal respiratory failure: A prospective randomized study. *Pediatrics* 76: 479–487.

Beauchamp, Tom L., and James F. Childress. 1994. *Principles of biomedical ethics*. 4th ed. Oxford: Oxford University Press.

Berry, Donald A. 1989. Comment: Ethics and ECMO. *Statistical Science* 4: 306–310.

Cowan, Dale H. 1981. The ethics of clinical trials of ineffective therapy. *IRB: A Review of Human Subjects Research* 3: 10–11.

Cummings, B.J. 1993. Anal cancer – radiation alone or with cytotoxic drugs? *International Journal of Radiation Oncology Biology Physics* 27: 173–175.

Cummings, B.J., A.R. Harwood, T.J. Keane, G.M. Thomas, and W.D. Rider. 1980. Combined treatment of squamous cell carcinoma of the anal canal: Radical radiation therapy with 5-fluorouracil and mitomycin-C, a preliminary report. *Diseases of the Colon and Rectum* 23: 389–391.

Dworetz, April R., Fernando R. Moya, Barbara Sabo, Igor Gladstone, and Ian Gross. 1989. Survival of infants with persistent pulmonary hypertension without extracorporeal membrane oxygenation. *Pediatrics* 84: 1–6.

Einhorn, Lawrence H., and John Donohue. 1977. Cis-diamminedichloroplatinum, vinblastine, and bleomycin combination chemotherapy in disseminated testicular cancer. *Annals of Internal Medicine* 87: 293–298.

Einhorn, Lawrence H., Stephen D. Williams, Michael Troner, Robert Birch, and Frank A. Greco. 1981. The role of maintenance therapy in disseminated testicular cancer. *New England Journal of Medicine* 305: 727–731.

Field, David. 1995. Ethics of consent for babies in randomised clinical trials. *Lancet* 345: 1370.

Flam, Marshall, Madhu John, Thomas F. Pajak, Nicholas Petrelli, Robert Myerson, Scotte Doggett, Jeanne Quivey, et al. 1996. Role of mitomycin in combination with fluorouracil and radiotherapy, and of salvage chemoradiation in the definitive nonsurgical treatment of epidermoid carcinoma of the anal canal: Results of a phase III randomized intergroup study. *Journal of Clinical Oncology* 14: 2527–2539.

Freedman, Benjamin. 1987. Equipoise and the ethics of clinical research. *New England Journal of Medicine* 317: 141–145.

Fried, Charles. 1974. *Medical experimentation: Personal integrity and social policy.* Amsterdam: North-Holland.

Glasziou, Paul, Iain Chalmers, Michael Rawlins, and Peter McCulloch. 2007. When are randomised trials unnecessary? Picking signal from noise. *British Medical Journal* 334: 349–351.

Golden, Gerald T., and J. Shelton Horsley III. 1976. Surgical management of epidermoid carcinoma of the anus. *American Journal of Surgery* 131: 275–280.

Goodman, Kenneth W. 2003. *Ethics and evidence-based medicine. Fallibility and responsibility in clinical science.* Cambridge: Cambridge University Press.

Green, Jerold P., Willis C. Schaupp, Simeon T. Cantril, and Gerald Schall. 1980. Anal carcinoma: Current therapeutic concepts. *American Journal of Surgery* 140: 151–155.

Greenhalgh, Trisha. 2010. *How to read a paper: The basics of evidence-based medicine.* 4th ed. Chichester: Wiley.

Gunderson, Leonard L., Kathryn A. Winter, Jaffer A. Ajani, John E. Pedersen, Jennifer Moughan, Al B. Benson III, Charles R. Thomas Jr., et al. 2012. Long-term update of US GI Intergroup RTOG 98-11 phase III trial for anal carcinoma: Survival, relapse, and colostomy failure with concurrent chemoradiation involving fluorouracil/mitomycin versus fluorouracil/cisplatin. *Journal of Clinical Oncology* 30: 4344–4351.

Hill, Austin Bradford. 1963. Medical ethics and controlled trials. *British Medical Journal* 1: 1043–1049.

Lantos, John D. 1997. Was the UK Collaborative ECMO trial ethical? *Paediatric and Perinatal Epidemiology* 11: 264–268.

Marwick, Charles. 1990. NIH 'Research Risks Office' reprimands hospital institutional review board. *Journal of the American Medical Association* 263: 2420.

McGlynn, Elizabeth A., Steven M. Asch, John Adams, Joan Keesey, Jennifer Hicks, Alison DeCristofaro, and Eve A. Kerr. 2003. The quality of health care delivered to adults in the United States. *New England Journal of Medicine* 348: 2635–2645.

Miller, Franklin G., and Steven Joffe. 2011. Equipoise and the dilemma of randomized clinical trials. *New England Journal of Medicine* 364: 476–480.

Newman, Howard K., and Stuart H.Q. Quan. 1976. Multi-modality therapy for epidermoid carcinoma of the anus. *Cancer* 37: 12–19.

Nigro, Norman D., V.K. Vaitkevicius, and Basil Considine Jr. 1974. Combined therapy for cancer of the anal canal: A preliminary report. *Diseases of the Colon and Rectum* 17: 354–356.

O'Rourke, P. Pearl, Robert K. Crone, Joseph P. Vacanti, James H. Ware, Craig W. Lillehei, Richard B. Parad, and Michael F. Epstein. 1989. Extracorporeal membrane oxygenation and conventional medical therapy in neonates with persistent pulmonary hypertension of the newborn: A prospective randomized study. *Pediatrics* 84: 957–963.

Pelligrino, Edmund D. 1979. *Humanism and the physician.* Knoxville: University of Tennessee Press.

Pocock, Stuart J. 1993. Statistical and ethical issues in monitoring clinical trials. *Statistics in Medicine* 12: 1459–1469.

Richie, Jerome P. 1998. Neoplasms of the testis. In *Campbell's urology,* ed. Patrick C. Walsh, Alan B. Retig, E. Darracott Vaughan Jr., and Alan J. Wein, 7th ed., 2411–2452. Philadelphia: W. B. Saunders.

Royall, Richard M. 1991. Ethics and statistics in randomized clinical trials. *Statistical Science* 6: 52–62.

Sackett, David L., William M.C. Rosenberg, J.A. Muir Gray, R. Brian Haynes, and W. Scott Richardson. 1996. Evidence based medicine: What it is and what it isn't. *British Medical Journal* 312: 71–72.

Samuels, Melvin L., Douglas E. Johnson, and Paul Y. Holoye. 1975. Continuous intravenous bleomycin (NSC-125066) therapy with vinblastine (NSC-49842) in stage III testicular neoplasia. *Cancer Chemotherapy Reports* 59: 563–570.

Schafer, Arthur. 1982. The ethics of the randomized clinical trial. *New England Journal of Medicine* 307: 719–724.

Schapira, David, and Alfonso Solimano. 1988. Is extra corporeal membrane oxygenation (ECMO) necessary to reduce mortality and morbidity in patients with meconium aspiration syndrome (MAS)? Abstract. *Pediatric Research* 23: 424A.

Sischy, Benjamin, John H. Remington, Sidney H. Sobel, and Edwin D. Savlov. 1980. Treatment of carcinoma of the rectum and squamous carcinoma of the anus by combination chemotherapy, radiotherapy and operation. *Surgery, Gynecology & Obstetrics* 151: 369–371.

Toomasian, John M., Sandy M. Snedecor, Richard G. Cornell, Robert E. Cilley, and Robert H. Bartlett. 1988. National experience with extracorporeal membrane oxygenation for newborn respiratory failure. Data from 715 cases. *ASAIO Transactions* 34: 140–147.

Truog, Robert D. 1993. Randomized controlled trials: Lessons from ECMO. *Clinical Research* 40: 519–527.

Tsimberidou, Apostolia-Maria, Fadi Braiteh, David J. Stewart, and Razelle Kurzrock. 2009. Ultimate fate of oncology drugs approved by the US Food and Drug Administration without a randomized trial. *Journal of Clinical Oncology* 27: 6243–6250.

Tukey, John W. 1977. Some thoughts on clinical trials, especially problems of multiplicity. *Science* 198: 679–684.

UK Collaborative ECMO Trial Group. 1996. UK collaborative randomised trial of neonatal extracorporeal membrane oxygenation. *Lancet* 348: 75–82.

Ware, James H., and Michael F. Epstein. 1985. Extracorporeal circulation in neonatal respiratory failure: A prospective randomized study. *Pediatrics* 76: 849–851.

Wei, L.J., and S. Durham. 1978. The randomized play-the-winner rule in medical trials. *Journal of the American Statistical Association* 73: 840–843.

Welton, Mark Lane, and Nalini Raju. 2011. Anal cancer. In *The ASCRS textbook of colon and rectal surgery*, ed. David E. Beck, Patricia L. Roberts, Theodore J. Saclarides, Anthony J. Senagore, Michael J. Stamos, and Steven D. Wexner, 2nd ed., 337–357. New York: Springer.

Williams, Stephen D., Robert Birch, Lawrence H. Einhorn, Linda Irwin, F. Anthony Greco, and Patrick J. Loehrer. 1987. Treatment of disseminated germ-cell tumors with cisplatin, bleomycin, and either vinblastine or etoposide. *New England Journal of Medicine* 316: 1435–1440.

Wilson, Stephen M., Oliver H. Beahrs, and Roberto Manson. 1976. Squamous cell carcinoma of the anus. *Surgery Annual* 8: 297–303.

Worrall, John. 2007. Evidence in medicine and evidence-based medicine. *Philosophy Compass* 2 (6): 981–1022.

———. 2008. Evidence and ethics in medicine. *Perspectives in Biology and Medicine* 51: 418–431.

Wung, Jen-Tien, L. Stanley James, Eitan Kilchevsky, and Edward James. 1985. Management of infants with severe respiratory failure and persistence of the fetal circulation, without hyperventilation. *Pediatrics* 76: 488–494.

Wyatt, John K., and Lloyd N. McAninch. 1967. A chemotherapeutic approach to advanced testicular carcinoma. *The Canadian Journal of Surgery* 10: 421–426.

Zelen, M. 1969. Play the winner rule and the controlled clinical trial. *American Statistical Association Journal* 64: 131–146.

Zelen, Marvin. 1979. A new design for randomized clinical trials. *New England Journal of Medicine* 300: 1242–1245.

Index

A
Accuracy, 4, 69–75, 79, 81–84, 90, 92–94, 96–98, 112, 122, 123
Achinstein, P., 2–4, 9, 19, 23–28, 32, 37, 91, 92, 94
Acute rheumatic fever, 57, 58, 78, 82, 96
Acute tubular necrosis (ATN), 60
Adair, F.E., 103, 104
Agent Orange, 47
Anal canal, cancer of, 139–141
Anderson, K.E., 55, 77, 88, 95
Argument
    bad lot, 31
    from indifference, 31
Atheromatous embolism (AE), 60, 74, 96

B
Background knowledge privilege, 31
Bacterial infection, 34, 37
Bartelink, H.F., 141
Bartlett, R.H., 130–132, 134–136
Bayes' theorem, 19, 20, 23, 60, 61, 74, 77–79, 89, 90, 110, 115, 117
Bayesianism, 2, 3, 19–23, 91, 94, 97
Beauchamp, T.L., 133, 138
Berghmans, R., 64
Berksonian bias, 70
Berkson, J., 70, 71
Berry, D.A., 90, 138
Bias
    Berksonian, 70
    information, 45, 54, 69, 81, 123
    obsequiousness, 42
    recall, 94

    selection, 69, 106
Birth control pills (BCPs), 15, 16
Blank, O., 75, 76, 80
Boden, W., 3
Breast, cancer of, 4, 102–108, 115, 117–120
British doctor study, 47, 79, 91, 95
Brown, B. Wm. Jr., 20, 89, 96
Butts, R., 7
Byar, D.P., 113

C
Caldwell, G.G., 70, 112
California class size study, 121
Cancer
    anal canal, 139–141
    breast, 73, 98, 102–108, 112, 117, 119, 121, 122, 124, 141
    lung, 47, 79, 98, 103, 141
    pancreas, 55, 79, 95
    testis, 5
    vagina, 50
Cancer of the anal canal, 5
Cannegieter, S.C., 81
Carnap, R., 13, 25
Cartwright, N., 68
Case-control study, 45–51, 55, 70, 72, 73, 82, 95, 96
Case reports, 4, 50, 62, 63, 98, 102, 124
Case studies, 4, 87, 94–97, 102
Casey, J.D., 56, 78, 81, 91, 96
Checkoway, H., 54
Childbed fever, 30
Childress, J.F., 133, 138
Christensen, D., 14

© Springer Nature Switzerland AG 2020
J. A. Pinkston, *Evidence and Hypothesis in Clinical Medical Science*, Synthese
Library 426, https://doi.org/10.1007/978-3-030-44270-5

Printed in the United States
by Baker & Taylor Publisher Services